International Series in Operations Research & Management Science

Volume 244

More information about this series at http://www.springer.com/series/6161

Thomas Hanne • Rolf Dornberger

Computational Intelligence in Logistics and Supply Chain Management

Springer

Thomas Hanne
Institute for Information Systems
University of Applied Sciences
and Arts Northwestern Switzerland
Olten, Switzerland

Rolf Dornberger
Institute for Information Systems
University of Applied Sciences
and Arts Northwestern Switzerland
Basel, Switzerland

ISSN 0884-8289 ISSN 2214-7934 (electronic)
International Series in Operations Research & Management Science
ISBN 978-3-319-40720-3 ISBN 978-3-319-40722-7 (eBook)
DOI 10.1007/978-3-319-40722-7

Library of Congress Control Number: 2016943140

Printed on acid-free paper

This Springer imprint is published by Springer Nature
The registered company is Springer International Publishing AG Switzerland

Preface

Over the last decades, logistics and supply chain management (SCM) have become one of the most often and intensively discussed fields in management and economics. Although many ideas and concepts used in logistics and SCM are reasonably old, much effort has been undertaken to transfer them into practice and to improve them further. Many publications, in academia as well as in application-oriented literature, have appeared. Logistics and SCM have become fields which are rich in terms of innovation and progress.

Despite these promising developments, there are still obstacles to bring advanced visions of improved planning and cooperation along logistics processes and supply chains into reality. On the one hand, there are many practical issues such as the availability and transparent processing of information, difficulties in establishing cooperation, or because of an increasingly uncertain or rapidly changing planning environment. On the other hand, it has become more and more apparent that the underlying planning problems are very complex and hard to solve even in the case that respective data is fully retrievable and complete.

From a computational point of view, many of these problems can be characterized as NP-hard, which means that the number of possible solutions is increasing exponentially with the problem size and that presumably no algorithms exist, which can solve them exactly within acceptable time limits—at least when the problems are "rather large." Unfortunately, most real-world problems can be considered rather large.

Especially during the last 20 years, these problems have been investigated intensively in the academic literature, and many suitable solution approaches have been suggested. As the problems usually cannot be solved exactly within an acceptable time, these methods allow to find sufficiently good, although not necessarily, optimal solutions.

One of the still growing streams of methods belongs to the field of computational intelligence (CI), which comprises mostly approaches inspired by concepts found in nature, e.g., the natural evolution or the behavior of swarms. These methods are based on general heuristic ideas and concepts for problem solving, which can—with some adaptations—be applied to a wide range of problems. To distinguish these methods from simple heuristics, which are often very specific to a single type of problem, they are also denoted as metaheuristics.

Although the respective computational intelligence methods have been studied in numerous applications related to logistics and supply chain management, they are hardly discussed in general textbooks in these fields. Often, the treatment of formal planning problems in these books does not go much beyond some rather simple and general results, which are often not applicable in real-world settings, for instance, the more than 100-year-old equation for calculating economic order quantities.

The book is intended to reduce this gap between general textbooks in logistics and supply chain management and recent research in formal planning problems and respective algorithms. It focuses on approaches from the area of computational intelligence and other metaheuristics for solving the complex operational and strategic problems in these fields.

Thus, the book is intended for readers who want to proceed from introductory texts about logistics and supply chain management to the scientific literature, which deals with the usage of advanced methods. For doing so, state-of-the-art descriptions of the corresponding problems and suitable methods for solving them are provided. The book mainly addresses students and practitioners as potential readers. It can be used as additional reference for undergraduate courses in logistics, supply chain management, operations research, or computational intelligence or as a main teaching reference for a corresponding postgraduate level course. Practitioners may read the book to become familiar with advanced methods that may be used in their area of work. For a reader, a basic understanding of mathematical notation and algebra is suggested as well as introductory knowledge on operations research (e.g., on the simplex algorithm or graphs).

The book is organized as follows: The first two chapters provide general introductions to logistics and supply chain management on the one hand and to computational intelligence on the other hand. The subsequent chapters cover specific fields in logistics and supply chain management, work out the most relevant problems found in those fields, and discuss approaches for solving them. In Chap. 3, problems in transportation planning such as different types of vehicle routing problems are considered. Chapter 4 discusses problems in the field of production and inventory management. Chapter 5 considers planning activities on a finer level of granularity, which is usually denoted as scheduling. While Chaps. 3 to 5 rather discuss planning problems, which appear on an operative level, Chap. 6 discusses the strategic problems with respect to the design of a supply chain or network. The final chapter provides an overview of academic and commercial software and information systems for the discussed applications.

We hope to provide the readers a comprehensive overview with specific details about using computational intelligence in logistics and supply chain management.

Olten, Switzerland — Thomas Hanne
Basel, Switzerland — Rolf Dornberger

Acknowledgments

The authors would like to express their gratitude to their employer, the University of Applied Sciences and Arts Northwestern Switzerland, particularly the School of Business, for supporting the proofreading of the book. Our dedicated thanks go to Christine Lorgé, assistant at our Institute for Information Systems, who read this rather scientific book about computational intelligence and logistics with great passion, although she is not coming from these disciplines.

Our deep gratitude goes to our beloved families, i.e., our wives and children. As professors who are active in research and teaching, with Rolf additionally being head of the institute and Thomas being head of one of its competence centers, we spend so much time with working issues that we always feel that our families are missing out. Therefore, we wish to express to them our highest thanks for their great understanding and their never-ending support! Thomas additionally thanks his wife Doris for proofreading some of the chapters, for discussion of some contents, and for support with the lists of symbols and acronyms.

Contents

List of Symbols

λ	Number of offspring (parameter of an evolution strategy)
π	Permutation
$\Re$	Set of real numbers
ρ	Number of recombined parents (parameter of an evolution strategy)
σ	Standard deviation, prediction error
σ_d	Standard deviation for demand
σ_{LT}	Standard deviation for lead time
μ	Number of parents (parameter of an evolution strategy)
A	(Feasible) set of alternatives, search space
A, B	Start and destination location of a transport
a	Capacity requirement per unit
a_i	Capacity requirement at location i
(a_{i1}, a_{i2})	Coordinates of customer i
a_{ij}	Influence factors (in particle swarm optimization)
a_p	Capacity requirement per unit of item p
at_i	Arrival time of job i
b	Capacity requirement per setup
C, C_i	Completion time (of job i)
c, c_t	Costs, unit costs
c_i^{best}, c_i^{global}	Acceleration factors (in particle swarm optimization)
c_{ij}	Transport costs (between i and j)
C_{max}	Makespan
D	Total required quantity (in the time horizon)
$d(.,.)$	Distance function
dd_i	Due date of job i
d_{ij}	Distance between i and j
dl_i	Deadline of job i
d_p	Demand of item p
d_t	Demand in t

$\hat{d}_t$	Predicted demand in t
$\overline{d}$	Average demand
E	Set of edges (arcs) of a graph
ea	Earliness
e_i	Edge of a graph
$f(.)$	Objective function
f_i	Finishing time
f_j	Setup costs of facility j
$G=(V,E)$	Graph
$G=(V,E,c)$	Graph with weights (which correspond to costs)
h, h_t, h_p	(Unit) holding costs
h^r	Holding cost rate
i_{pt}	Inventory of product p in t
i_{pt}^A, i_{pt}^B	Inventory of product p at location A (B) in t
i_t	Inventory in t
L	Set of locations
L^C	Set of customer locations
L^D	Set of depot locations
LT	Lead time
$\overline{LT}$	Average lead time
la	Lateness
l_i	Location
l_j^C	Customer location
l_i^D	Depot location
M	Large constant number
m	Number of neurons
$N(0, \sigma)$	Normal distribution with expected value 0 and standard deviation σ
n	Number of variables (e.g., locations in a tour)
$n!$	Factorial function of n
$O(.)$	Run time complexity of an algorithm (big O notation)
P	Subset of locations
p	Price per unit
p	Number of facilities to open in a p-median problem
p^{global}	Global best position of a particle (in particle swarm optimization)
p_i	Particle i (in particle swarm optimization)
p_i^{best}	Best previous position of particle i (in particle swarm optimization)
p_{ij}	Processing time of job i (on machine j)
q, q_j	Capacity (e.g., of a vehicle or a facility)
q_{ij}	Transported quantity between two locations
RP	Reorder point
r_i^{best}, r_i^{global}	Random coefficients (in particle swarm optimization)
S	Set of facility locations
SS	Safety stock
s_i	Start time of job i

s_{ik}	Arrival time of vehicle k at location i
s_i^{max}	Latest arrival time at location i (with specified time window)
s_i^{min}	Earliest arrival time at location i (with specified time window)
T	Planning horizon, time horizon, total number of periods
t_{ij}	Travel time from i to j
u, u_t, u_p	Fixed costs per order, setup costs of a production process
U_i	Order-up-to level quantity of product i
V	Set of vertices (nodes) of a graph, set of vehicles
v_i	Vertex (node) of a graph, velocity vector of a particle (in particle swarm optimization)
w_i	Inertia weight (in particle swarm optimization)
w_L, w_E	Weights (for lateness and earliness)
w_t	Production capacity in period t
x, x_{ij}, x_{ijk}, x_{pt}	Decision variables
x_o	Economic order quantity
x_t	Production quantities in t
y_t, y_{pt}	Binary decision variables
Z	Service level
z	Auxiliary variable for objective function values
z_{it}	Binary variables (denoting whether node i is visited at time t)

List of Abbreviations and Acronyms

2E-VRP	Two-echelon vehicle routing problem
ABC	Artificial bee colony
ACO	Ant colony optimization
AGV	Automated guided vehicles
AI	Artificial intelligence
AIMMS	Advanced interactive multidimensional modeling system
AIS	Artificial immune system
AMPL	A mathematical programming language
ANN	Artificial neural network
AP	Alternating position crossover
APS	Advanced planning systems or advanced planning and scheduling
AS/RS	Automated storage/retrieval system
ATP	Available-to-promise (functionality used in APS for supporting order promising and fulfillment)
BA	Bees algorithm
BE	Bionic engineering
BOM	Bill of materials
CCLSP	Coordinated capacitated lot-sizing problem
CFLP	Capacitated facility location problem
CI	Computational intelligence
CMA	Covariance matrix adaption
CMA-ES	Evolution strategy with covariance matrix adaptation
CMWP	Capacitated multi-facility Weber problem
COI	Cube-per-order index
CS	Cuckoo search
CSCMP	Council of supply chain management professionals
CULSP	Coordinated uncapacitated lot-sizing problem
CVRP	Capacitated vehicle routing problem
CX	Cycle crossover

DE	Differential evolution
DLL	Dynamic linked library
DPSO	Discrete particle swarm optimization
DPX	Distance-preserving crossover
EA	Evolutionary algorithm
EC	Evolutionary computation
ELS	Economic lot size
EOQ	Economic order quantity
EP	Evolutionary programming
ER	Genetic edge recombination crossover
ERP	Enterprise resource planning
ERX	Edge recombination crossover
ES	Evolution strategy
FA	Firefly algorithm
FIFO	First-in first-out
FL	Fuzzy logic
FNN	Feedforward neural network
GA	Genetic algorithm
GAMS	General algebraic modeling system
GCLSP	General capacitated lot-sizing problem
GDP	Gross domestic product
GE	Grammatical evolution
GEP	Gene expression programming
GIS	Geographic information systems
GNX	Generalized n-point crossover
GP	Genetic programming
GPS	Global Positioning System
GRASP	Greedy randomized adaptive search procedure
GSO	Glowworm swarm optimization
HNN	Hopfield neural networks
HS	Harmony search
IEEE	Institute of Electrical and Electronics Engineers
ILS	Iterated local search
ILS-FDD	Iterated local search with fitness-distance-based diversification
JiT	Just-in-time
LCS	Learning classifier systems
LGP	Linear genetic programming
LOX	Linear order crossover
LSAP	Linear sum assignment problem
MA	Memetic algorithms
MACS	Multiple ant colony system
MCFP	Multicommodity-flow problem
MCLC	Maximal covering location criterion
MGP	Metagenetic programming

MICLSP	Multi-item capacitated lot-sizing problem
MIP	Mixed-integer optimization problems
ML	Machine learning
MLLSP	Multilevel lot-sizing problem
MLULSP	Multilevel uncapacitated lot-sizing problem
MNS	Multi-neighborhood search
MOO	Multiobjective optimization
MPX	Maximal preservative crossover
MRP	Material requirement planning
MRP II	Manufacturing resource planning
NC	Natural computing
NN	Neural networks
NP	Nondeterministic polynomial (complexity class)
NP-hard	Complexity class
NSGA, NSGA-II	Non-dominated sorting genetic algorithms
OOP	Object-oriented programming
OpenOPAL	Software toolbox for optimization and learning
OPL	Optimization programming language
OptimJ	Java-based modeling language for optimization
OX, OX1, OX2	Order (or order-based) crossover
PACO	Population-based ant colony optimization
PMX	Partially mapped crossover
PNN	Perceptron neural networks
POS	Position-based crossover
POX	Precedence preserving order-based crossover
PPS	Precedence preserving shift mutation
PSO	Particle swarm optimization
RFID	Radio-frequency identification
RIL	Reinforcement learning
ROI	Return-on-investment
RVNS-VNS	Reduced and standard variable neighborhood search
SA	Swarm algorithms
SC	Soft computing
SCM	Supply chain management
SCX	Sequential constructive crossover
SI	Swarm intelligence
SICLSP	Single-item capacitated lot-sizing problem
SLAP	Stock location assignment problem
SLSP	Single link shipping problem
SOFM	Self-organizing feature maps
SOM	Self-organizing maps

SPEA, SPEA2	Strength Pareto evolutionary algorithms
SPT	Shortest processing time
SSCFLP	Single source capacitated facility location problem
TMS	Transportation management system
TSP	Traveling salesman problem
TSPLIB	Traveling salesman problem library
UFLP	Uncapacitated facility location problem
UMWP	Uncapacitated multi-facility Weber problem
USILSP	Uncapacitated single item lot-sizing problem
VLSN	Very large-scale neighborhood search
VND	Variable neighborhood descent
VNDS	Variable neighborhood decomposition search
VNS	Variable neighborhood search
VR	Voting recombination crossover
VRP	Vehicle routing problem
VRPTW	Vehicle routing problem with time windows
WES	Warehouse execution system
WMS	Warehouse management system
ZIO	Zero inventory ordering

Chapter 1
Introduction to Logistics and Supply Chain Management

Abstract In this chapter we provide a brief introduction into the concepts of logistics and supply chain management. Considering different production factors, functions and processes in a company and across companies both terms are specified taking into account different definitions from the literature. After that a brief reflection of logistics history is given, followed by a discussion of the modern importance of logistics and supply chain management. The last section motivates the usage of advanced planning methods as discussed in later chapters of the book.

1.1 The Concept of Logistics and Supply Chain Management

Roughly simplified a company can be considered as a system which receives a specific input, creates goods or services out of it and delivers the output to its customers. These three main activities can be denoted as procurement (or purchasing), production, and distribution (or sales). Very often input and output are physical objects or materials which require a specific handling such as transportation or storage in order to realize the basic functions of a company. These activities are considered the core tasks of logistics (Fig. 1.1).

To be more precise, the input of a company, often called the production factors, can be distinguished into input which is used up during the production and input which is available for a longer period of time. The first group of production factors is often just called materials or consumption factors, whereas the second group includes the classical production factors capital and labor (employees). These factors are also often called resources and include machines, equipment, and buildings. They are not used up during the production but frequently their capacity (available time over a period) needs to be matched with the time needed for production. Also "modern" production factors like information, human capital or management belong to the latter group.

In a more traditional understanding logistics only refers to the materials and could therefore also be called material logistics. Of course, many concepts from

T. Hanne, R. Dornberger, *Computational Intelligence in Logistics and Supply Chain Management*, International Series in Operations Research & Management Science 244,
DOI 10.1007/978-3-319-40722-7_1

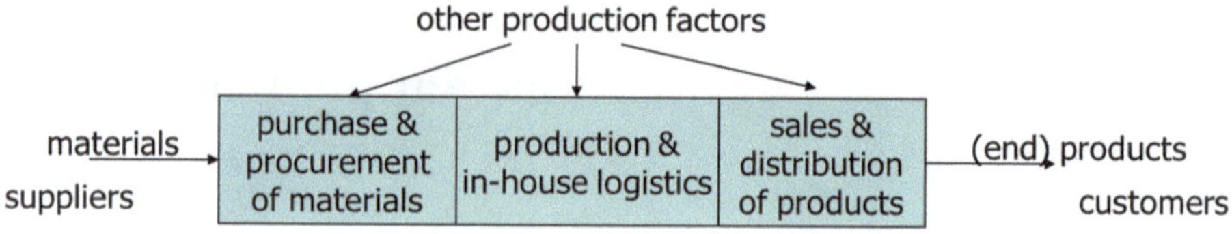

Fig. 1.1 The company and its main functions

logistics can be applied to other "objects" like human beings as well. Humans (e.g. tourists) too need to be transported and accommodated.

Expressed in a more abstract way, logistics deals with transfers of materials in space, time, and quantity from the procurement of materials needed for production via the storage of materials, intermediate products, and finished products to the physical distribution to customers. Thus, logistics focusses on the planning and execution of spatial, temporal and quantitative transfers.

Spatial transfers could simply be called transportation and can be distinguished into long- and short-distance transports. Long distance transportation means transports between different locations such as warehouses, plants, and different companies that primarily use trucks, trains, ships, and aircrafts. Short-distance transportation means transports inside a location (plant, warehouse). This type of spatial transfer is occasionally also called material flow. In the short-distance transportation usually different devices are used than in long-distance transportation, e.g. fork lifts, conveyors, or automated guided vehicles.

Temporal transfer means "transport" over time, i.e. from today when a material is available to the future when the material is needed. This is the purpose of what we call more simply storage or warehousing.

Quantitative transfers take place when, for instance, large amounts of some goods are provided in smaller quantities. This is one of the usual activities of retailers which buy goods in larger quantities from the producers or wholesalers and usually sell them in smaller amounts to end customers. Changes of quantity also take place when customer orders are fulfilled. Ordered items are picked from the warehouse (where they are usually available in larger quantities), brought together, packaged and sent to the customers.

Taking these three aspects into account, we can define the main tasks of logistics as processes for the settlement of differences in space, time and quantity of goods. Let us note that production itself is not considered a logistics activity. Logistics is rather everything that is needed "around" the production with respect to the physical products which are finally provided to customers.

Another frequently used possibility to define logistics is to express the tasks of logistics by different aspects that must be done right. Today this is specified by the six (or seven) Rs of logistics: Have the **r**ight items (material), in the **r**ight quantity, at the **r**ight time, at the **r**ight place, in the **r**ight quality (condition), with the **r**ight costs (and the **r**ight information).

If we look at more recent definitions of logistics, we can observe that they have been formulated in a more elaborate way. For instance, Christopher (2010, p. 4) defines logistics as "the process of strategically managing the procurement, movement and storage of materials, parts and finished inventory (and the related information flows) through the organization and its marketing channels in such a way that current and future profitability are maximized through the cost-effective fulfillment of orders."

Another rather general definition of logistics management by the Council of Logistics Management describes logistics as "the process of planning, implementing, and controlling the efficient, effective flow and storage of goods, services, and related information from point of origin to point of consumption for the purpose of conforming to customer requirements" (Lambert 2008).

Thus logistics does not only concern the production of physical goods but goods and services in general. It is therefore also relevant for public administration and institutions like hospitals, schools and service-providing companies like traders, banks and other financial service providers or insurance companies. Another interesting aspect in such definitions is that logistics does not only refer to in-house processes in a company and processes with direct market partners but includes processes beyond that scope. Strictly speaking, "from point of origin to point of consumption" includes everything from the initial production of agricultural products or the extraction of raw materials in mines to finished products used by customers (or companies). Usually, the relating processes involve a large number of companies and it is often neither useful nor possible to consider every step of these processes. Realistically, only those parts of the overall processes should be taken into account where a common planning of the activities is possible and makes sense for the involved partners.

Summarizing, there is a rather narrow point of view which limits logistics to the transport and storage of physical goods (material logistics) and a wider point of view which includes immaterial goods (services), considers neighboring processes and extends the scope to other companies in the supply and demand network. We chose a wider point of view but consider mostly situations in material logistics as they are more illustrative.

The trend towards wider definitions of logistics has been incorporated in the idea of what is today called supply chain management. In particular, this concept has been motivated by the awareness that many logistic processes are not just relevant in a considered company but that such processes should also be considered at suppliers and customers for providing a good product or service to end customers. Moreover, from the viewpoint of involved companies, it often makes sense to plan these activities in an integrated way to recover the contribution of the partners in the supply network and to maximize their added value.

Let us have a look at some recent definitions of Supply Chain Management. Cooper and Ellram (1993) define it as "an integrative philosophy to manage the total flow of a distribution channel from the supplier to the ultimate user". Harland (1996) defines supply chain management as "the management of a network of interconnected businesses involved in the ultimate provision of product and service packages required by end customers". A more official definition comes from the Council of Supply Chain Management Professionals (CSCMP): "Supply chain management encompasses the planning and management of all activities involved in sourcing and procurement, conversion, and all logistics management activities. Importantly, it also includes coordination and collaboration with channel partners, which can be suppliers, intermediaries, third party service providers, and customers. In essence, supply chain management integrates supply and demand management within and across companies."

1.2 A Short History of Logistics

The origin of the word "logistics" is quite old. It is derived from the old Greek word "logos" which means something like (written) speech, language, word, reason, ratio, and calculation. Apart from maybe "ratio" there is little obvious connection to our modern day understanding of logistics.

However, the aspect of calculation was considered for denoting an administrative person in ancient Rome or Greece as "Logistika" who was responsible for finance, procurement, and distribution, mainly with a military focus (Tudor 2012). Such a focus on military activities was taken up in modern times. During the nineteenth century, the terms tactics, strategy and logistics were broadly used to distinguish essential activities for being successful in warfare. According to the Oxford English Dictionary logistics was "the branch of military science relating to procuring, maintaining and transporting material, personnel and facilities." Thus, it was not military activity in a closer sense but everything what was identified as necessary for mastering such activities.

Just like tactics and strategy, logistics found its way from military planning into the civil sector. The first wider usage in business took place during the 1960s in the U.S., mainly with the focus on planning and organizing distribution activities.

From then on logistics became more and more popular, its meaning and areas of application were extended significantly. Logistics was supplemented by additional concepts, in particular "supply chain management" which came up in the 1980s and focused on an even wider field. The insight that logistics should not just focus on isolated activities (such as a single transport or the warehousing of a good at a defined location) was the starting point of a process-oriented thinking and the consideration of logistic networks (or "supply chains"). It became evident that logistic activities are usually closely connected with other (logistic or non-logistic) activities so that an integrated planning can produce a higher utility for the customers or a greater added value for the involved companies.

Although it seems that logistics and supply chain management are rather modern ideas, there is much evidence that even complex logistics problems have been successfully dealt with in the distant past. If we go back to prehistoric times, the earliest human civilizations were denoted as hunters and gatherers (or hunter-gatherers). Thus, these civilizations were named after a logistic activity, the procurement of food. Apart from that, we can assume that other logistic activities played a major role as well: Due to the varying supply of food—often uncertain and fluctuating during a year—measures for storing food were essential for the survival of a tribe. Another aspect was the distribution of food: Since not all members of a tribe went hunting or gathering food—and since the hunters and gatherers were not all equally successful—it made sense to share the prey and the harvest for a better division of labor and a reduction of risks.

Another prehistoric example of logistics is the transport and trade of goods such as stones, pottery, metals, or other raw materials sometimes even over long distances. For instance, there is good evidence that Obsidian, a volcanic glass with a limited occurrence throughout the world, was transported over several hundred kilometers (Dogan and Michailidou 2008).

Ancient examples for large logistics operations are, of course, antique wars of conquest. Impressive examples in the civil sector include, e.g. water irrigation systems, which are sometimes based on networks of canals, tunnels or pipes covering some hundred kilometers, for instance in the Maya, the Persian or the Roman civilization.

1.3 Recent Trends and the Modern Importance of Logistics

Although many problems and concepts in logistics are not that new, there are good reasons why logistics has become more important during the last decades and why the increase of public (and academic) attention in this area is not just a modern hype.

On the one hand, there are a number of general economic trends which facilitate an increased engagement into logistics questions. On the other hand, technological progress was significant in logistics and related fields.

Let us start with general economic aspects. One of the long-term concepts for discussing economic development is the three-sector hypothesis. Roughly speaking, this hypothesis is based on a classification of economic activities into three sectors: The primary sector which deals with the production of raw materials (especially agriculture), the secondary sector which includes manufacturing and industry, and the third or tertiary sector which deals with services. During the maturing of economies usually the following development takes place: First (i.e. some hundred years ago) the primary sector strongly dominates. Then the secondary sector grows and becomes the most important, e.g. during the industrial revolution in Europe and the U.S. in the nineteenth century. Later on and in particular in the industrialized countries during the second half of the twentieth century, the second sector starts to shrink while the third sector grows and, finally, dominates. These long-term developments demonstrate the importance of logistics because logistic activities can be classified as services and belong, therefore, to the growing third sector.

Although this says little about the growing importance of particular logistics activities, there is further evidence: Along with the long-term economic development, growth and increased standards of living can be observed. These increased standards of living lead to more complex and challenging customer needs. For instance, customers require a higher quality of goods and services, the speed and the security of deliveries have become more important, and special requests like individualized goods or services are often to be taken into account as well. Some of these aspects have clear logistic implications: For a fast and secure supply, good forecasts of demand, an adequate warehousing or fast transportation systems are very important. Other aspects such as the quality of products are partially dependent on a good mastering of logistics. The selection of suppliers, an efficient handling of materials, the avoidance of deficient products, an adequate dealing with customer claims, etc. may contribute to the quality perceived by customers.

Apart from such customer-centric aspects various other global developments in the economy are relevant for the evolution of logistics and supply chain management: One of them is the economic liberalization. Over a longer period of time we have observed a strengthening of free trade. Since Ricardo's proof of the advantages of free trade in the nineteenth century, tolls and customs duties were reduced or abolished in many parts of the world. Also other restrictions on international trade, such as import or export contingents were reduced in many cases. The freedom of establishment was realized in many parts of the world and it has become much easier to set up a business in other countries today. The entrance to specific industries and markets (e.g. in the shipping trade) was disburdened. All this leads to specific opportunities but also to new challenges due to an increased competitive pressure.

Ever since their fall at the end of the 1980s, the previously communist countries have played a special role. Today, most of them have transformed into market economies. Other nations having been called developing countries only a few decades ago have shown a strong economic growth, and some of them are rather called emerging markets today. Famous among them are the BRIC countries Brazil, Russia, India, and China. This group of countries was first analyzed by O'Neill (2001) and later on expected to become larger (in terms of GDP) than the G6, the formerly largest industrialized countries (USA, Japan, Germany, France, UK, and Italy) before 2050 (Wilson and Purushothaman 2003).

Especially since the beginning of the twenty-first century trade between distant parts of the world (especially between East Asia and Europa/U.S.) has increased significantly. Whereas countries like China were broadly appraised to produce cheap low-quality goods 10 or 20 years ago, today many high quality products are manufactured there. The competitive challenges are evident. Not only production conditions in different parts of the world become more similar, but customer desires assimilate as well. For instance, eating Chinese food in Western countries became quite usual already some decades ago. But also a reverse trend can be observed meanwhile: While a product like chocolate played an infinitesimal role in China 20 years ago, it has become popular today. So there are also clear opportunities for industries in Western countries which are often focusing on rather expensive high-quality products. Rising incomes in emerging countries make products more affordable for the population.

As mentioned before, a certain consequence of this globalization is that more and more goods are transported over longer distances. As a result, increased transportation costs in the economies can be expected. If we look at the transportation costs (e.g. for the U.S., see below) in percentage of the GDP we observe that they rather decreased over the last decades. This is even more remarkable because the prices for fuel tended to increase over that time (with strong fluctuations). The main reasons for this can be found in aspects relating to technical progress. First of all, many types of vehicles (trucks, planes, vessels) are more fuel-efficient than before. Often, vehicles have become larger (especially ships) which also leads to lower costs per ton transported cargo. Economies of scale do not just arise because of technical reasons. For instance, the crew for a larger ship can be kept constant in size.

Along with such aspects of single transportation activities the whole infrastructure for transports has been upgraded. For instance, turnover activities in a harbor can be done more smoothly today. The container as a well standardized transport equipment plays a major role. A transport chain comprising truck, train or ship transportation (intermodal transport) works more efficiently than in pre-container times. Transportation is nowadays better supported by information and communication technology than in the past. GPS-based navigation systems facilitate the planning and execution of transports. Better planning algorithms allow for more efficient transportation. This aspect will be treated in more details in Chap. 3.

Apart from these transportation related aspects, the technical progress in information and communication technologies supports other logistics activities as well (and, of course, business activities in general). Some of the most obvious aspects have been the improved performance of computers and the decrease of respective costs during the last decades. Another obvious development is the rise of the Internet and the emergence of e-business. Let us just consider one example of how logistics is directly influenced by these trends: Because of e-business today goods are more often directly shipped to customers instead of being distributed through retailers. Compared to shipments to retailers this requires a significantly larger number of usually smaller shipments.

Another less visible effect of the technological progress is the increased usage of automation technology in companies. Today, companies often master their in-house logistic tasks by using a significant amount of material flow technology or robotics. Examples of such technologies are conveyors for in-house transportation, automatic storage and retrieval systems, automated guided vehicles (AGVs), or an increased usage of mobile devices by humans. One specific technology which has been discussed intensively during the last 10 years is RFID (radio-frequency identification). Small chips (or smart tags) which consist of an electronic circuit (including a memory and possibly sensors) and an antenna for communication can be attached to goods (or parcels). Such technology can be used for a better identification of objects, for a better tracing and tracking during their transportation, or for more intelligent purposes such as the monitoring of a cold chain for frozen food.

Table 1.1 Logistics costs in the US

	Total logistics cost (%)	Inventory and related costs (%)	Transportation costs (%)
1981	16.2	8.3	7.3
1986	11.6	4.9	6.3
1991	10.6	4.3	5.9
1996	10.3	3.9	6
2001	9.5	3.4	5.8
2003	8.6	2.8	5.4
2006	9.9	3.2	5.8
2007	10.1	3.5	6.2
2008	9.4	2.9	6.1
2009	7.7	2.5	4.9
2010	8.3	2.7	5.2
2011	8.5	2.8	5.3
2012	8.5	2.8	5.3
2013	8.5	2.9	5.3
2014	8.0	2.7	5.0

Source: Own computations and Wilson (2013)

Along with new planning technologies and respective software, such technical progress can lead to innovative logistics solutions and/or the decrease of costs.

If we want to understand the present importance of logistics, it makes sense to consider the quantitative figures of that economic sector. For the US there is a regular statistics about logistics in the annual "US State of Logistics Reports" which are commissioned by the Council of Supply Chain Management Professionals (CSCMP). For 2014 the following data for logistics costs are shown in the respective report (Wilson 2013; Robinson 2015):

- 917 Bio. US$ for transportation (702 Bio. US$ for motor carriers)
- 476 Bio. US$ for warehousing/storage including interest costs and taxes, etc.
- 56 Bio. US$ for logistics administration
- 1449 Bio. US$ total logistics costs

When seeing these impressive numbers, we should bear in mind that many costs for logistics services (esp. those provided internally) are not even considered in the official statistics. Very often, companies do not even know their own logistics costs, because parts of them are not well attributed in their financial accounting (e.g. costs for in-house transportation).

Table 1.1 shows the logistics costs for the US expressed in percentage of the gross domestic product (GDP) of this country. Total logistics costs and the two main subcategories, inventory-related costs and transportation-related costs, are depicted (The category administrative costs is omitted so that the two percentages of the subcategories are slightly less than the percentage for the total logistics costs.) Moreover, the developments of these costs from the early 1980s until 2012 are shown.

When looking at the temporal development it is interesting to see that the percentage declined over the last 30 years (with a few years showing a re-increase of the costs). The biggest cost reductions were obtained in the first half of the 1980s, around 2000, and then again after 2007.

The percentages for the subcategories show that the biggest decreases were realized for the inventory-related costs, from 8.3 % in 1981 to 2.8 % in 2012. On the one hand, this can be attributed to some general economic reasons such as a poor economic situation by the beginning of the 1980s which often goes along with high inventories. On the other hand, we can assume that progress was made by an increased awareness of this cost-component in companies, by a stronger focus on supply chain management concepts, and by using related planning techniques. Moreover, we can assume that the cost savings were not reached at the expense of more frequent stock-out situations. Stock-outs were reduced over the last 40 years but then seemed to persist at an average level of about 7–8 % in the retail business (Fernie and Grant 2008). In their extensive study, Gruen et al. (2002) find a world-wide level of about 8.3 % stock-outs being somewhat higher in Europe and lower in the USA and speculate about being a "natural level" for stock-outs.

Not only the inventory-related logistics costs were reduced, but also some reduction in the transportation costs can be observed, in particular during the 1980s and after 2008. This is astonishing because the fuel prices, a major component of the transportation cost, had a tendency to increase—although with strong fluctuations. This reduction of transportation costs is even more notable because we find an increase of shipping volumes and an increase of transportation distances during the last decades. We assume, therefore, that the prevalent reasons for the decrease of transportation costs (in percentage of the GDP) are the technological progress and the usage of better planning techniques.

The logistics cost data for the US seem to be comparable to those from other parts of the world. Table 1.2 shows absolute logistics costs (measured turnover from logistics service providers) and logistics costs in percentage of the GDP for several European countries (Klaus and Kille 2009; DHL without year). We see that the costs are mostly in the range of 7–9 % with some exceptions, Italy with a significantly lower percentage and Finland with a much larger percentage.

1.4 The Need for a Better Planning

As we have seen, the importance of logistics has grown but at the same time the competition and the cost pressure were enormous and led to decreased amounts of money spent for logistics activities. There is reason to assume that this trend will hold up in the future. A major challenge is therefore how companies can gain strategic competitive advantages concerning logistics activities.

Such competitive advantages can be manifold, but cost aspects will probably play a dominant role. Other competitive advantages or specific objectives related to

Table 1.2 Logistics costs in Europe

	Turnover in logistics (Bio EUR in 2007)	Logistic costs (in % GDP 2007)
Germany	205	7.8
France	111.9	8.5
UK	107.3	8.3
Italy	81.3	5.3
Spain	81	8.3
Netherlands	45.9	8.4
Belgium	31.8	7.2
Sweden	28	8.6
Poland	26.1	(missing)
Finland	22.4	14.7

logistics can be, in particular, aspects related to time, aspects related to quality, and aspects related to flexibility.

With respect to time shorter delivery times and quicker responses to changes in demand are frequently discussed. Because of technological progress or volatile customer preference, the time to market of new products and services also becomes increasingly important.

Quality relates to many aspects which contribute to the satisfaction of customer requirements. They relate to the physical products but also to accompanying services e.g. delivery times, avoidance of scrap, or adequate reactions to customer complaints and a quick and satisfactory reconditioning of products. Flexibility is more difficult to specify but focusses on a company's potential for fast and appropriate reactions to changes in customer preferences and demand, competition, and other environmental factors.

Many different management activities—depending on the specific industries and markets—can contribute to reaching these goals. In any case, a better planning supports the achievement of such goals. Impressive examples of the impact of a better planning, especially a better planning based on advanced quantitative methods, can be found on the INFORMS website "Getting Started with Analytics" (INFORMS 2016). The website is connected with a database including dozens of brief case studies showing the potentials of analytics in a diverse range of industries, functional areas, or with respect to different kinds of benefit. A significant amount of the included cases is related to logistics aspects such as inventory management, transportation, routing or scheduling, or deals with supply chain management in a more general sense. The reported benefits are often cost savings (or increased revenues) in the range of several dozens of Mio US$ up to more than 100 Mio US$. In his survey on several success stories, Lustig (2013) reports cost savings often in the range of 5 % or more and can report on savings for particular resources of 30 % or more when respective planning techniques were used.

For such reasons, there is a strong need for a better planning of logistics activities for reducing costs and reaching other goals. Some of the progress in logistics during the last decades has already been achieved by a better planning of the respective processes. For some processes, the usage of state-of-the-art planning technologies has already become reality. For instance, when a single transportation is to be planned, the usage of navigation software is already common practice. Such tools allow for an easy and precise location of the current position of a person, a good, or a vehicle. They provide current information with respect to maps and transportation networks. And they allow for the calculation of a rather precise shortest path (or quickest path) for going from A to B. Other planning areas where a lot of progress has taken place are, for instance, the forecast of demand or the planning of inventory levels.

Nevertheless, more difficult planning tasks are not yet supported very well in practice. This has to do with the fact that many of the related optimization problems are computationally hard so that efficient algorithms for solving them to optimality do not exist. Moreover, some of the related formal problems are rather new so that not much effort has been made to provide practical solutions and software.

Another difficulty is that frequently the logistics problems of a company are hard to standardize. A general formulation of the planning problem and a design and implementation of the respective software is inadequate for the specific tasks. This absence of a "general problem solver" makes it more difficult and costly to develop and provide solutions which are adequate for the actual planning problems of a company. From a more theoretical point of view, we know that often the devil is in the details. Minor changes in a problem formulation may have a major impact on its complexity or the suitable algorithms to solve it.

In this book we are trying to formulate typical planning problems with respect to logistics, and present algorithms which are basically suitable for solving these problems. As a specific problem is usually more or less unique, particular adaptations of problem formulations or specifications of algorithms appear to be necessary. Fortunately, the class of methods considered, i.e. approaches from computational intelligence and in particular metaheuristics, are adaptable. They are designed for a problem-specific adaptation. Although, many of them can be specified by generic formulations, usually their power and superiority compared with classical optimization approaches requires a problem-specific design. This concerns, for instance, the question of how a problem (and its solutions) should be encoded to make it treatable for a respective method. Moreover, the methods usually consist of basic steps or phases (often called operators) which also allow for problem-specific adaptations. For instance, an evolutionary algorithm is based on several evolutionary operators such as mutation, recombination, or selection which can be implemented in numerous ways. Each of them can be or should be specific for being most effective for the considered optimization problem. Later on in this book we will consider several of such adaptations.

References

Cooper, M. C., & Ellram, L. M. (1993). Characteristics of supply chain management and the implications for purchasing and logistics strategy. *International Journal of Logistics Management, 4*(2), 13–24.

Christopher, M. (2010). *Logistics and supply chain management* (4th ed.). Harlow: Financial Times/Prentice Hall.

DHL. (without year). *The macroeconomic significance of logistics*. Accessed March 12, 2016, from http://www.dhl-discoverlogistics.com/cms/en/course/trends/macroeconomics.jsp

Dogan, I. B., & Michailidou, A. (2008). Trading in prehistory and protohistory: Perspectives from the eastern Aegean and beyond. In C. Papageorgiadou & A. Giannikouri (Eds.), *MELETIMATA 53: Sailing in the Aegean, readings on the economy and trade routes* (pp. 17–53). Athens: Institute of Greek and Roman Antiquity (IGRA), National Hellenic Research Foundation.

Fernie, J., & Grant, D. B. (2008). On-shelf availability: The case of a UK grocery retailer. *International Journal of Logistics Management, 19*(3), 293–308.

Gruen, T. W., Corsten, D. S., & Bharadwaj, S. (2002). *Retail out-of-stocks: A worldwide examination of extent. causes and consumer responses*. Washington: Grocery Manufacturers of America.

Harland, C. M. (1996). Supply chain management, purchasing and supply management, logistics, vertical integration, materials management and supply chain dynamics. In N. Slack (Ed.), *Blackwell encyclopedic dictionary of operations management*. Oxford, UK: Blackwell.

INFORMS. (2016). *Getting started with analytics*. Accessed March 12, 2016, from https://www.informs.org/Sites/Getting-Started-With-Analytics

Klaus, P., & Kille, C. (2009). *Der deutsche Logistikmarkt bleibt weiterhin ein Wachstumsmarkt! Deutliche Zuwächse bei Mengen und Beschäftigung*. Hamburg: Fraunhofer Institut Integrierte Schaltungen, Deutscher Verkehrs-Verlag.

Lambert, D. M. (2008). *Supply chain management: Processes, partnerships, performance* (3rd ed.). Ponte Vedra Beach, FL: Supply Chain Management Institute.

Lustig, I. (2013). *Optimization-based solutions: Smarter decisions for a smarter planet*. IBM Research – Business Analytics and Mathematical Sciences. Accessed March 12, 2016, from https://irinadumitrescublog.files.wordpress.com/2013/04/optimization-smarter-planet-irv-public.pdf

O'Neill, J. (2001). *Building better global economic BRICs*. Global Economics Paper No: 66, New York: Goldman Sachs.

Robinson, A. (2015). *A complete breakdown of the 26th State of Logistics Report*. Accessed March 12, 2016, from http://cerasis.com/2015/07/10/state-of-logistics-report/

Tudor, F. (2012). Historical evolution of logistics. *Revista de Ştiinţe Politice/Revue des Sciences Politiques, 36*, 22–32.

Wilson, D., & Purushothaman, R. (2003). Dreaming with BRICs: The path to 2050. *Global Economics Paper*, 99. New York: Goldman Sachs.

Wilson, R. (2013). *24th Annual State of Logistics Report*. Lombard, IL: Council of Supply Chain Management Professionals (CSCMP).

Chapter 2
Computational Intelligence

Abstract This chapter provides an introduction to Computational Intelligence (CI). Artificial Intelligence (AI) and CI are briefly compared. CI by itself is an umbrella term covering different branches of methods most of them following the paradigm "nature-inspired". While CI and AI partly overlap, the methods applied in CI benefit from nature-inspired strategies and implement them as computational algorithms. Mathematical optimization is shortly explained. CI comprises five main branches: Evolutionary Computation (EC), Swarm Intelligence (SI), Neural Networks, Fuzzy Logic, and Artificial Immune Systems. A focus is laid on EC and SI as the most prominent CI methods used in logistics and supply chain management. EC is coupled with Evolutionary Algorithms (EA). Methods belonging to EC respectively EA are Evolution Strategy, Genetic Algorithm (GA), Genetic and Evolutionary Programming, the multiobjective variants Non-dominated Sorting GA (NSGA) and Strength Pareto EA (SPEA), Memetic Algorithms, and further methods. The most important methods belonging to SI are Particle Swarm Optimization (PSO), Discrete PSO and Ant Colony Optimization. EA and SI approaches are also attributed to the class of metaheuristics which use general problem-solving concepts for problem solution during their search for better solutions in a wide range of application domains.

2.1 Foundations of Computational Intelligence

For hundreds of millions of years, nature has continuously developed a bunch of techniques to overcome obstacles in the daily lives of plants, animals and humans and has found solutions to our problems within the real world.

T. Hanne, R. Dornberger, *Computational Intelligence in Logistics and Supply Chain Management*, International Series in Operations Research & Management Science 244,
DOI 10.1007/978-3-319-40722-7_2

2.1.1 Artificial and Computational Intelligence and Related Techniques

2.1.1.1 Artificial Intelligence

Artificial Intelligence (McCarthy 2007), abbreviated AI, was "scientifically born" in the middle of the last century. Around 1955, John McCarthy (McCarthy et al. 1955) was the first scientist to introduce AI as a technology in computer science, stating that AI was the basis for a "truly intelligent machine". The interest in AI as a branch of computer science has been rapidly growing since. Today, many other disciplines such as engineering, biology, psychology, linguistics, etc. are involved in or concerned with AI in an interdisciplinary way.

While research in AI is still growing, definitions, classification and wording become blurred (Neapolitan and Jiang 2012): Originally, it was distinguished between *Strong AI* and standard AI (also called normal AI or *Weak AI*). Strong AI means general intelligence in the sense of the creation of human-like intelligence of machines. Thus, strong AI focusses on the development of intelligent machines (i.e. software or hardware robots) and the design of human cognition comprising creativity, consciousness and emotions to make the robots somehow human-like. The aim here is to match or even to exceed human intelligence. In contrast, weak AI is more dedicated to the solution of particular problems applying human-like approaches. Reasoning, adaptation, learning and social intelligence are key components, which are used in computer programs in order to solve particular problems (e.g. text or speech recognition).

In other words: Weak AI aims to provide intelligent algorithms somewhere stored inside the software, whereas strong AI yields rather an embodiment of intelligence in robot-like machines and makes them behave somehow intelligent. Today, the two directions of AI are mixing again: AI is understood as making machines intelligent as well as creating intelligent machines.

AI comprises various methods for the intelligent processing of information (e.g. perception, learning, planning, knowledge, reasoning, and communication) and provides—particularly in robotics—machines and computers with the ability to move in, interact with, and participate in the real world. Therefore, AI mostly uses symbolic techniques such as logic- and knowledge-based methods, decision trees, case-based reasoning, and stochastic automata. With the help of these techniques, logical conclusions or automata states are derived.

Many research directions in AI have been evolved over the last 60 years: Probabilistic methods are used for uncertain reasoning. Classifiers and statistical learning methods support the learning. Search and optimization methods perform mathematical optimization. Neural networks, logic and control theory have emerged. Programming languages for artificial intelligence (e.g. Lisp and PROLOG) were developed. And recently, the research in intelligent machines and robotics has been making big leaps.

But, in spite of all its success AI has a disadvantage, namely the lack of more general fault-tolerant mechanisms: AI is often "too logical" to robustly solve problems that have not been well-posed, as is the case with most real-world problems. But, reality is the perfect test bed. Reality is imperfect/not perfect, often uncertain, noisy and volatile—reality is real. This is how our world looks and behaves. This is where Computational Intelligence (CI) plays an important role.

2.1.1.2 Computational Intelligence

Computational Intelligence (Kruse et al. 2013) (abbreviated CI) is often mentioned in relation with AI. Many researchers state (Fulcher and Jain 2008), that they do not refer to CI as a sub-part of AI: but that CI and AI comprise partly the same or similar methods. However, CI and AI use different methodologies and approaches to describe the principles of their methods. The name CI emerged in the 1990s (Poole et al. 1998), and first focused on intelligent robots. Today, interestingly, CI seems to be more clearly defined as AI: CI mainly deals with "nature-inspired" computational methods for facilitating intelligent behavior in complex and possibly changing environments (Engelbrecht 2007).

Furthermore, CI differs from AI in the sense that CI uses (mostly) sub-symbolic techniques, that is to say techniques that go beyond a certain symbolic level. A state or a problem instance is expressed or represented by numerical numbers instead of a symbolic entity as generally done in AI. Depending on the application areas and the particular problems, the advantages of CI methods might be superior to AI: CI methods perform efficient approximations within reasonable computation time because they often use non-deterministic, stochastic components, heuristics and metaheuristics, which converge quickly and robustly to one or more solutions even in the presence of noise, uncertainty, moving targets and changing search spaces.

Other advantages of CI methods are that they often work efficiently only needing a rough model of the problem, that they are fault-tolerant, and that they are mostly well-suited for parallelization. On the other hand, the biggest disadvantage of CI, compared to deterministic methods, is that the optimal solution(s) can generally be neither proved nor guaranteed to be found.

2.1.1.3 Techniques Related to Artificial and Computational Intelligence

Further related areas of research are *Machine Learning* (Mitchell 1997) (ML), *Soft Computing* (SC), *Natural Computing* (Rozenberg et al. 2012) (or Natural Computation, NC), *Bionics* (or bionic engineering, BE), and many more.

ML (Machine Learning) is counted as a sub-category of AI and deals with learning from data. Its relation to CI is only given by the neural networks. Furthermore, the paradigms of ML do not necessarily focus on "nature-inspired".

SC (Soft Computing) is an area of research which investigates methods to efficiently (but inexactly) find solutions to computational extensive and hard problems in huge search spaces. NC (Natural Computing) combines the paradigm "nature-inspired" with the categories "computing natural phenomena" and often "employing natural materials". Thus, NC focusses on the synthesis of nature by means of computing, on the development of novel nature-inspired hardware and on the understanding of natural phenomena as information processing.

CI, SC and NC comprise partly the same methods, such as neural networks, evolutionary computation, swarm intelligence and fuzzy logic. But the application focus of CI is to solve various real-world problems, whereas that of SC is to compute hard problems in huge search spaces, and the application focus of NC is to understand nature by computing and rebuilding it.

Another related discipline is BE (bionics engineering). In contrast to the other mentioned research areas, focusing on computer science aspects, bionics plays a role on the physical or technical level. In bionics, the aim is to imitate the biological structures or physical abilities of fauna and flora and to transfer these abilities into technical products. Nevertheless, bionics has contributed to the research of evolutionary computation, neural networks, and swarm intelligence by imitating mechanisms or processes found in nature. Examples of bionicly engineered products are augmented material surfaces to reduce the drag in fluid dynamics (e.g. shark skin, bird wings) or to avoid the adherence of dirt (e.g. lotus effect).

2.1.1.4 Interest in Computational Intelligence

The potential of CI is big; the interest in CI is great and still growing. Increasingly more conferences deal with CI. The most important ones are probably the bi-annual IEEE World Congress on Computational Intelligence (IEEE WCCI) with the annual conferences IEEE Congress on Evolutionary Computation (IEEE CEC), the International Joint Conference on Neural Network (IJCNN), and the IEEE International Conference on Fuzzy Systems (FUZZ-IEEE), as well as the annual IEEE Symposium Series on Computational Intelligence (SSCI). The largest CI organization is the IEEE Computational Intelligence Society, which initiates these conferences.

2.1.2 Properties of Computational Intelligence

The main paradigm of CI is "nature-inspired". This comprises various facets: Biology and particularly flora and fauna provide different mechanisms in nature, which have been researched and transferred to algorithms mimicking these mechanisms on a computational level. Such mechanisms are for example the evolution of living things, which succeeded over many generations to adapt better to their environmental conditions and to be superior to the others. Further mechanisms are the simulation of the functionality of brains to learn and remember, or the reaction to the intrusion of pathogens into the body, or the adaption to the behavior of swarms like birds, fish or ants do.

Most methods of CI are more or less derived from nature. Nature is providing particular problem-solving strategies in the real-world, particularly how to best live, survive, and reproduce in certain environments. While CI exclusively covers such methods, AI knows only some of the nature-inspired methods (e.g. neural networks as the most common method applied in both, AI and CI).

Investigating nature, the methods of CI are inspired by biological models or processes. They are cast into theoretical models and then coded in algorithms, often without adapting them specifically onto the application area. Typically, methods of CI can be characterized by these three aspects:

- They are nature-inspired computational methodologies.
- They solve complex problems without (much) problem-specific knowledge.
- They are well suited to solve real-world problems.

Today, CI is categorized within the following five main branches. We try to characterize each CI branch featuring three individual properties:

- Evolutionary Computation (EC):
  - EC applies biological mechanisms of evolution by using the principle of survival of the fittest.
  - A population, consisting of several individuals, will improve over generations by selection, crossover and mutation.
  - EC is powerful in solving optimization and search problems within non-uniform search spaces.
- Swarm Intelligence (SI):
  - SI is the collective behavior of self-organizing multi-agent systems.
  - The entirety of a population consisting of simple agents, who are interacting only locally, leads to an intelligent-like global behavior.
  - SI mostly helps to solve optimization and search problems, often belonging to a kind of distance minimization.
- *Neural Network* (NN; respectively Artificial Neural Network, ANN)
  - NNs are inspired by real biological neural networks of brains of animals or human beings.
  - Data are processed through a network of interconnected artificial neurons.
  - NNs identify relationships between input and output data, finding patterns and generating information.
- Fuzzy Logic (FL):
  - FL is a many-valued logic, similar to the human reasoning.
  - It allows the usage of approximate values and incomplete or ambiguous data.
  - FL provides approximate solutions or conclusions.

- Artificial Immune System (AIS):
  - AIS is inspired by the immune system of humans, animals and plants.
  - It adapts the characteristics and processes of their immune system for learning and memory.
  - AIS is applied in adaptive systems for problem solving.

Additionally, further methods are sometimes counted as branches of CI, sometimes not, such as *Reinforcement Learning* (Kramer 2009) and *Simulated Annealing*. Reinforcement Learning adapts the behavior of an artificial agent (e.g. a software agent or a robot) by rewarding wanted and penalizing unwanted behavior. Simulated Annealing uses a thermodynamic principle in physics: In annealing processes (i.e. metal cooling processes) the material tries to harden to a state of minimal energy. Simulating this effect on a computational level, simulated annealing is mainly used for minimization problems.

The methods of CI allow another kind of information processing as known from AI. CI methods adopt some simple principles of nature which allow them to search either for alternative solutions (e.g. one or more local minima or maxima) or global optima. Some of these methods are well suited for adaptation to changing environments or conditions—and are thus used for solving control problems.

Some of the CI methods are able to learn and to remember—consequently they are used in pattern recognition or classification. CI methods are more or less fault-tolerant against incomplete and noisy or indefinite inputs. Many CI methods allow parallel processing (e.g. EC and SI use sets of parallel available solution candidates). Often it is sufficient to apply the CI methods to an easy model. Thus, (almost) no problem-specific knowledge is needed. Overall, depending on the problem, CI often allows to apply more sophisticated, nature-inspired solution strategies than AI.

2.1.3 The Big Picture of Computational Intelligence

Figure 2.1 provides a systematic overview of CI—to be more specific a big picture of CI—with a focus on Evolutionary Computation (EC) and Swarm Intelligence (SI). In general, EC uses Evolutionary Algorithms (EA). Their main representative methods are Genetic Algorithms (GA), Evolution Strategies (ES), Genetic Programming (GP), and Evolutionary Programming (EP). Further methods belonging to EC and EA are Memetic Algorithms (MA), Learning Classifier Systems (LCS), Differential Evolution (DE), Harmony Search (HS), and some more.

In the past, Swarm Algorithms (SA) were often mentioned as a sub-category of EC as well. Nowadays, SAs are counted as representative methods of SI, which include, for instance, Ant Colony Optimization (ACO), Particle Swarm Optimization (PSO), Bees Algorithm (BA) and Cuckoo Search (CS). PSO knows further sub-methods such as binary, discrete or combinatorial PSO.

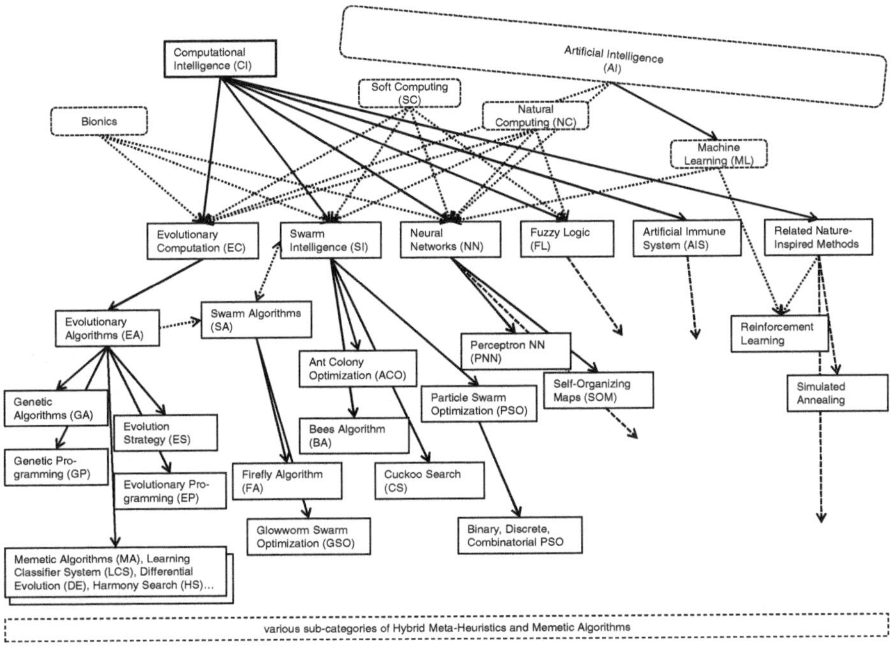

Fig. 2.1 Methods belonging to Computational Intelligence—with focus on the sub-branches of Evolutionary Computation and Swarm Intelligence

Today, the list of such nature-inspired algorithms has become quite long. For instance in the survey by Fister Jr et al. (2013) more than 70 methods are classified, most of them coming from the field of SI approaches. The search and optimization strategies from the field of EC and SI are also denoted as metaheuristics. But, some of them include other general problem-solving approaches, not necessarily referred to the paradigm “nature-inspired”.

The world of Neural Networks (NN) is even larger, where Perceptron Neural Networks (PNN) (also denoted as Feedforward Neural Networks or FNN) and Self-Organizing Maps (SOM) are the most prominent methods of NN. Fuzzy Logic (FL) and Artificial Immune Systems (AIS) are the other main CI branches. The branches NN, FL and AIS are only shortly mentioned. For, the main focus in our book is laid on EC and SI, because these respective methods are especially successful in logistics applications. At the end of this chapter we also discuss briefly a few further metaheuristics, which share similarities with EC and SI methods and also belong to the most frequently used approaches for logistics problems.

2.1.4 *Application Areas of Computational Intelligence*

2.1.4.1 Optimization and Search

A large, perhaps the largest, application area of CI is optimization in all its facets (Burke and Kendall 2006). In optimization, respectively mathematical optimization, the problem is—generally speaking—to find minima or maxima, which are possibly subject to constraints.

An optimization problem is mathematically defined as follows:

Given: A function $f : A \rightarrow \boldsymbol{R}$ from some set A to the real numbers
Sought: An element x_0 in A such that $f(x_0) \leq f(x)$ for all x in A (so-called *minimization*) or such that $f(x_0) \leq f(x)$ for all x in A (so-called *maximization*).

Mathematical definitions are well suited to clearly and uniquely explain a mathematical aspect. However, for non-mathematicians, these definitions are often not self-explaining. Therefore, we explain optimization to our students as follows, using the illustration shown in Fig. 2.2.

An obviously identifiable continuous optimization problem is to find the highest peak in a mountain range: The borderline between the mountains and the sky is comparable to the mathematical function $f(x)$ denoting the height of the mountains at every position x. The domain A is the search space (also called choice set) where all possible positions x are forming the set of candidate solutions (feasible solutions). A candidate solution x is thus a member of the search space A of possible solutions to a given problem.

The function $f(x)$ is the objective function, which is sometimes called cost function (in minimization problems) or utility function (in maximization problems) or *fitness function* if an evolutionary computation method is applied.

The optimization goal is to find the specific feasible solution x_0 that maximizes the objective function. This particular feasible solution x_0 with $f(x_0) \geq f(x)$ for all

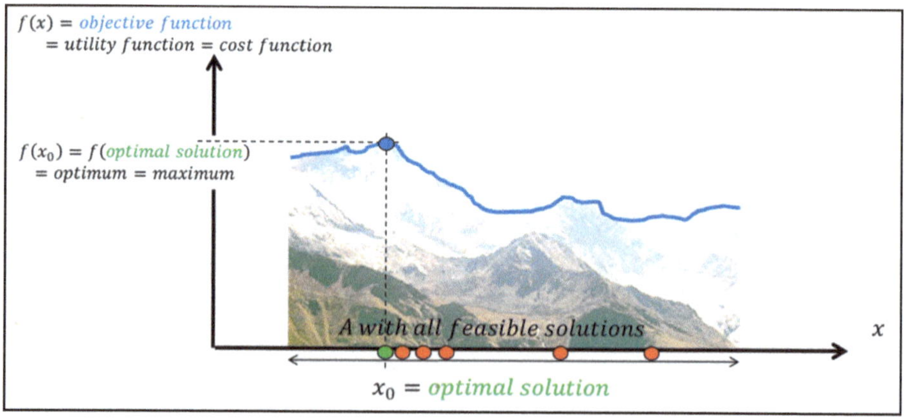

Fig. 2.2 An optimal solution x_0 and its maximum $f(x_0)$ of the objective function $f(x)$

x in A is then called the optimal solution. In minimization problems, the goal is to find $f(x_0) \geq f(x)$.

Additionally, an optimization problem often involves constraints, which limit the search space. Mathematically, the candidate solutions have to additionally satisfy inequality and equality constraints, such as $g(x) \leq 0$ and $h(x) = 0$.

Optimization problems can comprise continuous variables x or discrete x or combined ones in the search space. The cost function may be continuous or discrete and might exist only implicitly. Some optimization problems are combinatorial optimization problems, where a certain sequence of possible entities has to be found. Different ways to deal with such optimization problems are shown later.

In physics or engineering, optimization is often related to the maximization of force, power, efficiency or stability, or to the minimization of energy consumption, loss, time or faults. In economics, optimization tends to maximize profit, return-on-investment (ROI), economical gain, whereas minimization yields the smallest amount of invested resources. E.g. in logistics, optimization often yields the shortest, fastest or cheapest transportation route.

Related application areas are *search* problems which sometimes cannot be described as *optimization* problems. In search problems, the objective is to find either an item with specified properties among a set of other items, or even to find a feasible solution, or to find a diverse set of good alternatives (e.g. sets of local minima or maxima). Therefore, additional add-ons to the algorithms (such as niching or sharing) may extend the classical optimization algorithms.

The methods of CI do not make a real distinction between optimization and search. Depending on the characteristics of the problem (e.g. NP-hard) and on the search space (e.g. discrete, continuous) some CI methods are more suitable than others.

In general, the methods of CI are heuristics or metaheuristics: An approach is called a *heuristic*, if it is used to obtain high-quality solutions, although optimality cannot be guaranteed without checking all the possible solutions. *Metaheuristics* are based on general heuristic concepts. They can usually be applied to a wide area of different search or optimization problems, especially in computationally hard domains, where exact optimization approaches sometimes appear as insufficient or too time-consuming. Although convergence verifications (i.e. proofs for obtaining optimal solutions under suitable conditions) for some of the well-known metaheuristics are available, the notion of "metaheuristics" has been kept.

Some search methods or algorithms perform a deterministic search. This means that they always return the same answer, given exactly the same input and starting conditions. But, CI methods are (mostly) non-deterministic search methods or algorithms. This means that they may follow different optimization paths and return different answers given exactly the same problem. However, the different answers should (but need not) converge in the course of time to equally good solutions.

Furthermore, CI methods can do more than just mathematically search and optimize some variables: Some of these methods provide the ability to learn and classify, e.g. as used in data mining (e.g. particularly NN, AIS). Some are used to recognize patterns, in printed or handwritten texts as well as in spoken language (e.g. NN). Others are used to adapt to changing environments or conditions, a quality which is needed in designing controllers (e.g. EC, FL). Some help model and simulate e.g. biological evolution processes, brain functionality or technical or economical processes (e.g. EC, SI, NN). Some enable evolutionary design and arts or the composition of music (e.g. EC). Some are used in computer games or in agent based systems (e.g. SI, EC, NN), to model and simulate experiments which cannot be tested in reality, to process big data and so on.

2.1.4.2 Multiobjective Optimization

If more than one objective function has to be optimized mutually, the problem becomes a *multiobjective optimization* (MOO) problem. Multiobjective optimization, or multicriteria or multiattribute optimization, is the process of simultaneously optimizing two or more conflicting objectives, subject to certain constraints.

Generally, in such multiobjective problems one cannot identify a single solution that simultaneously optimizes each objective. While searching for solutions, the attempt to improve an objective further must lead to downgrade another objective. The result is a set of solutions, which are called non-dominated, Pareto optimal, or Pareto efficient. They are characterized by the fact that they cannot be neglected by replacing them with other solutions which improve one objective function without worsening another. Solving a multiobjective optimization problem means to find such non-dominated solutions, the Pareto optimal solutions.

In the section *Multiobjective Evolutionary Algorithms*, two evolutionary algorithms which perform a so-called Pareto optimization are presented.

2.2 Methods of Computational Intelligence

2.2.1 Evolutionary Computation

Evolutionary Computation (EC) goes back to the 1950s and 1960s, where new schemes of optimization via self-adaptation were investigated. At that time, EC was counted as a sub-field of AI. Today, EC is noted as a sub-field of CI as well, because the methods belonging to EC are adapting biological evolutionary mechanisms. Many expressions of the biological evolution have been used to describe the computational algorithms of EC:

- In general, the EC methods work with a set of solution candidates. Such a set is called a population. The solution candidates are referred to as individuals. The individuals are often encoded by so-called chromosomes, just like in nature.
- The EC methods are iterative methods. Within one loop, the set of individuals is called a generation. Thus, in EC the solution is computed by using a population of changing individuals, each of them representing a solution candidate.

- The EC methods iteratively improve the "quality" of the individuals. This quality is called fitness and is measured by the objective function, called the fitness function. While fitness often implies that the fitness has to be maximized, in minimization problems the objective function might be named as cost function.
- The iterative steps are generations where—from one generation to the next—some parent individuals pass on their good features and traits to the child individuals using mechanisms of reproduction such as crossover and recombination, mutation and selection.

The methods of EC are well suited to solve optimization and search problems just like evolution continuously optimizes the living beings in specific environments. These real-life environments are the search spaces in the mathematical optimization. If the search spaces change over time, the optimization process has to follow the moving targets. EC does so automatically due to its iterative character. This can be used in the development of controllers (e.g. driverless autonomous cars or control of the combustion process in burners (Büche et al. 2002)). If we understand the optimization process of EC as a design process, many applications can be found in design (e.g. development of gas turbines (Dornberger et al. 2000), the Mars Rover antenna, and computer circuits). Beyond mathematical optimization, natural sciences and engineering, the methods of EC are used in economics, social sciences and art.

2.2.2 *Evolutionary Algorithms*

Evolutionary Algorithms (EA) are either considered a sub-category of EC, or a synonym of EC. In general, EAs describe a set of stochastic, generic population-based, (meta)heuristic optimization algorithms. All the algorithms belonging to EA are characterized by reproduction applying selection, recombination, and mutation. Applying these operators leads to an artificial evolution process on a computational level mimicking the real evolution in particular aspects.

The most important methods that belong to EA are Genetic Algorithms (GA), Evolution Strategies (ES), Genetic Programming (GP), and Evolutionary Programming (EP). Surprisingly, Swarm Algorithms (SA) are sometimes also considered belonging to EA; but we link the SA to the other independent big branch of CI named Swarm Intelligence (SI). Additionally, all these methods provide different variants which result in many further metaheuristics: Differential Evolution (DE), Learning Classifier Systems (LCS), Harmony Search (HS) etc.

Hybrid metaheuristics result by merging two or more methods. Memetic algorithms (MA) were developed by combining EA methods with local search methods. Metaheuristics, which are hybridized with approaches from mathematical programming (such as branch-and-bound, the Simplex algorithm or Newton gradient methods), are usually denoted as *matheuristics*.

The algorithms defer by the representation of the problem, i.e. the coding of the solution candidates, by different working modes of the EA operators and by the order of the execution of these operators. Here, recombination does not necessarily need to take place, because one parent changed by mutations might be sufficient.

The principles of the EA methods are simple. The following list shows how narrow the relation between certain aspects in nature or biology and the computational algorithms of EA are:

- The paradigm taken from evolution in nature is Darwin's principle "survival of the fittest".
- In nature, biology encodes the properties of an organism, called phenotype, using four amino acids. On the computational level, this limitation does not exist; various, totally different encodings of the candidate solutions are possible on the level of the individuals. In general, the coding takes place with binary, integer or real values or with a set of given objects carrying the hereditary information, called genotype.
- Every individual, in the EA every encoded candidate solution, obtains a particular quality (called fitness) showing how good it is in solving the problem—in relation to the other candidate solutions.
- Mimicking the natural selection process, the better individuals are selected with higher probability in order to produce the children (new candidate solutions) for the next generation (next iteration).
- The selected individuals are recombined, where mutation additionally adds genetic variations.

In computer science, these principles are implemented into an algorithm. The sequence and importance of the single principles might differ from case to case. But, the general flowchart of an EA can be noted as follows and shown in Fig. 2.3 (left):

1. *A random initial set of individuals as zero (initial) generation is generated.*
2. *The fitness of each individual is computed.*
3. *While no termination criterion is fulfilled, a next new generation is generated.*
4. *Therefore, particular individuals, depending on their fitness, are selected in order to form the new set of individuals in the new generation.*
5. *(Within a certain probability) they are recombined using crossover.*
6. *(Within a certain probability) they are mutated.*
7. *Their new fitness is computed.*

The class diagram of an object-oriented programming (OOP) implementation of an EA is shown in Fig. 2.3 (right):

- The class Population contains *1-to-many* Generation(s). All Generations belong to *this* Population.
- Each Generation consists of *1-to-many* Individual(s), where these Individuals belong to *this* Generation.
- Each Individual has exactly *1* Fitness value.

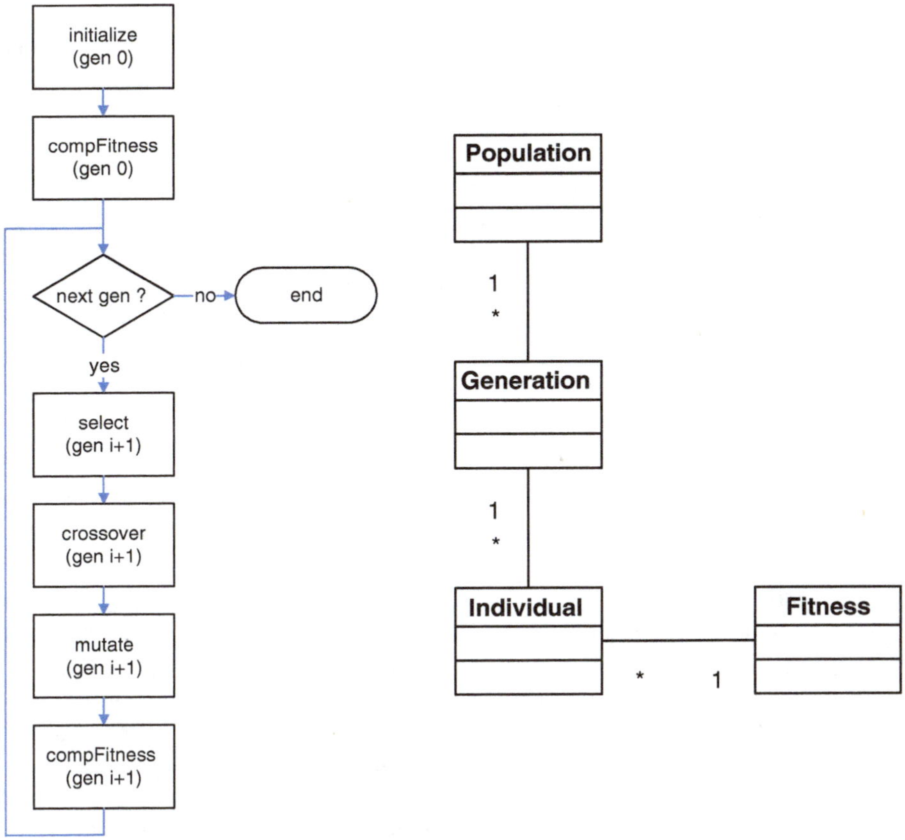

Fig. 2.3 General flowchart of an EA (*left*) and the class diagram of an object-oriented programming (OOP) implementation of an EA (*right*)

The selection defines which individuals form the population of the next generation. It can be used in different ways: Either those parent individuals are selected which are further used for recombination and mutation. Or a selection is used to take out the best individuals of all newly generated individuals and the existing old ones to form the next generation. Or a selection defines which individuals are eliminated and will not be used for the offspring. Hence, different selection operators are possible:

- The *standard selection* takes the n fittest, ranked individuals as parents, where n is the number of individuals per generation.
- *Roulette-wheel selection* allows a fitness-proportionate selection, where the possibility of choosing an individual depends on the magnitude of its fitness value in relation to the fitness values of the other individuals.
- *Tournament selection* takes the best individual of two randomly chosen parent individuals.

- *Truncation selection* takes a certain portion of the best individuals.
- *Elitism selection* (shortly called *elitism*) always copies the best individual from the old generation into the new generation in order to guarantee that no setback in the evolution process happens.

Different selection operators can be combined. Particularly, elitism is used frequently in order to always preserve the last best solution.

The advantages of the EA methods are that they are (almost) problem-independent and very robust. Thus, EA can be used as black-box optimization and search algorithms. (Almost) no knowledge about the problem which should be solved and only a slight knowledge about the search space are necessary. Various classes of optimization problems can be solved, because the problems must neither be continuous nor differentiable. The EA is a stochastic exploration of the search space for local and global optimization, and—in the ideal case—tremendously faster than an exhaustive search.

2.2.2.1 Evolution Strategy

In the 1960s and 1970s, Ingo Rechenberg (1994) introduced new methods for optimizing engineering designs, such as aerofoils or airplane structures. He called this approach *Evolution Strategy* (Rechenberg 1973) (ES), because a kind of artificial evolution was performed: The features of the better designs were combined, stochastically slightly changed and passed on to a next generation of modified designs. Although this approach was carried out on hardware designs, it is obvious that principles belonging to evolutionary computation were performed. Rechenberg combined this approach with bionics, transferring mechanisms from nature to technical respectively engineering products.

While Rechenberg at the beginning applied the ES rather in experimental optimization on hardware designs, Hans-Paul Schwefel (1977) extended the ideas more to computational optimization, and encoded the technical designs (phenotypes) using vectors with the real numbers from R^n (genotypes).

The principles of the ES for computational optimization are the following: A set of vectors consisting of n real-valued components represents a set of different solution candidates. Each vector component describes a particular feature and can be mutated by adding a normally distributed random value. The magnitude of this random value is called the mutation strength and is determined by certain adaptation mechanisms. Each vector is tested to solve the problem, and ranked depending on its fitness value.

Different main types of the ES with particular (μ, λ)-notation or $(\mu + \lambda)$-notation are possible. μ stands for the number of recent candidate solutions in the parent population, and λ for the number of additional candidate solutions generated from the parent population:

- A $(1 + 1)$-ES is the simplest ES. This configuration consists of one parent vector (candidate solution) and one modified vector. The better candidate solution is taken as parent for the next generation. The simplest ES corresponds to the greedy hill climbing algorithm: In each iteration, the greedy algorithm takes the actual best candidate solution out of a set of at least two alternatives.

- More common is the *(1 + λ)*-ES, which consists of one parent vector and λ modified ones. Here, the λ modified candidate solutions plus the recent parent solution compete to be selected for the next generation.
- In the *(1, λ)*-ES, only the λ modified vectors compete to be selected for the next generation.
- In order to overcome local optima, the *(μ, λ)*-ES proposes a generation which consists of μ parent vectors and λ modified ones, where (if $\lambda > \mu$) the μ best candidate solutions of the new λ ones are taken for the next generation.
- In the *(μ + λ)*-ES, all μ parent and λ modified vectors compete in the selection process for the next generation.
- Often, a recombination between the parent vectors is additionally applied, denoted by ρ and *(μ/ρ, λ)*-ES. ρ is the number of parents who contribute to the next generation using particular recombination operators.

In ES, mutation is the driving operator to modify the vectors representing the candidate solutions. However, to find the optimal magnitude of mutation, called the mutation strength, it is an optimization problem by itself. Rechenberg introduced the so-called 1/5-rule, which proposes to adjust the mutation strength in such a way that in each generation around 20 % of the generated candidate solutions are better than the original parent ones.

Another method to find a suitable mutation strength is self-adaptation, which integrates an additional component into the vector containing the mutation strength value. Thus, the mutation strength is optimizing itself.

A further extension of the original ES is the controlled mutation of vector components by a so-called covariance matrix adaptation (CMA). This kind of ES is called the CMA-ES (Hansen and Ostermeier 2001), which guides the mutation strength in the direction of the possibly highest increase of the fitness value. Although ES are stochastic derivative-free methods, the CMA-ES works similarly to the usage of derivative information in gradient methods.

2.2.2.2 Genetic Algorithm

In the 1970s, John H. Holland (1975) developed a method for solving search and optimization problems, applying the principles of EA. He called the method *Genetic Algorithm* (GA) and the candidate solutions *individuals*. These individuals are encoded by *chromosomes*, which are typically implemented as vectors or strings. Each component of the string (or vector) is called a gene. The main difference to the real-valued ES is that Holland introduced the simple binary coding $\{0, 1\}^N$ for encoding a solution instance.

While all variants of ES work rather on phenotype level, the variants of GA fully optimize on genotype level which is mapped from the real problem instances down

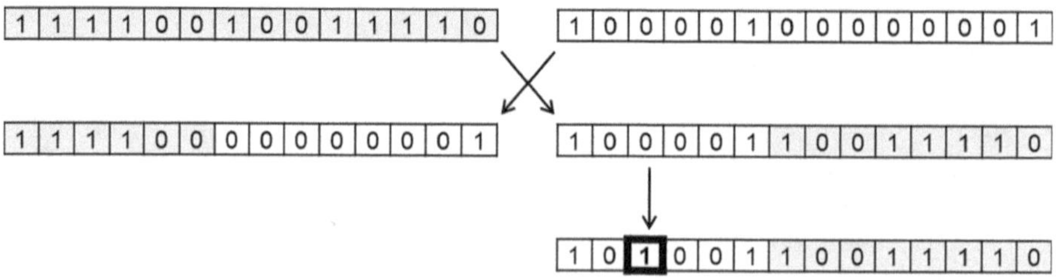

Fig. 2.4 Simple crossover between 6th and 7th gene applied to the two parent individuals (*top*) resulting in the two child individuals (*middle*). Additionally, the right child individual is mutated at the 3rd gene (*bottom right*)

to binary digits. To move to this level of encoded individuals, a unique mapping from the representation of the search space (e.g. real valued or integer discrete) is necessary. This is called a genotype-phenotype mapping. Holland's basic idea was that a representation with binary digits would be the easiest representation for a computer to handle.

Furthermore, the two values (*0* and *1*) of GAs are similar to the four values of the nucleic acid in nature (base-4 chromosomes of the DNA). However, the advantages of the original GAs with a binary coding diminished when computational power was increasing. Nowadays, various encoding schemes are used in GAs. A variable length coding of the chromosomes is also possible.

In a GA the genetic operators and the computation of the fitness work in the following way:

- *Recombination/crossover*: Recombination was originally introduced as a simple crossover, where two parent individuals exchange their genes at a certain crossover position, as shown in Fig. 2.4. Besides a *one-point-crossover*, an *n-point-crossover* exchanges sequences of genes between two individuals. In combinatorial problems, special types of crossover, such as partially mapped crossover (PMX), allow to recombine individuals' information without producing (too much) child individuals with invalid coding.
- *Mutation*: Using the binary coding, a mutation is obvious, because the bit of the particular gene has only to be inverted (Fig. 2.4). With the other coding schemes, a mutation has to be defined as how the value of a gene is changing to which other value.
- *Fitness*: The computation of the fitness value needs a remapping from the genotype representation to the phenotype level, because the problem that must be solved is often not represented in a binary search space. Furthermore, GAs can be interactive in the sense that the evaluation of the fitness value happens outside the GAs—perhaps interactively by the user or within a technical unit.

Holland and Goldberg (Goldberg 1989) describe the ability of GAs to converge to optima with the building block hypothesis. It says that a GA is able to collect positive traits stored in short sequences of genes, called building blocks, which are inherited and summed up over generations of improving individuals.

2.2.2.3 Genetic Programming

Genetic Programming (GP) was elaborated by John R. Koza (1992) in the 1980s to be applied for function approximation and for the automatic generation of computer code and programs. GP is based on GA and uses a genotype representation of the problem, e.g. of the mathematical formulae or of the entire source code. While GA mostly uses fixed-length individuals, GP allows individuals of non-fixed length.

In the standard GP, the chromosomes are tree-like representations of the mathematical functions or code expressions, as shown in Fig. 2.5. Using such trees, GP aims to optimize the entire structure of the tree rather than to optimize solely the values or operators of certain nodes.

The initial idea of GP prefers programming languages which naturally use tree-based internal data structures such as Lisp. Each chromosome is a code candidate representing a tree. Then, the nodes—and thus all branches depending on that node—are free to be exchanged, recombined and mutated. This happens not only between different chromosomes, but also within each particular chromosome. A side effect is that many code candidate solutions with an invalid syntax are generated, and have to be sorted out. The fitness is evaluated by the fulfilment of the syntax rules, the requirements of the developed code, and its "elegance", or rather its "parsimony".

There are many extensions of the original GP: In order to also allow other programming languages (beyond tree-based languages), such as the assembly language, or imperative ones, GP has been extended to *Linear Genetic Programming* (LGP). LGP codes the instructions used in the programming languages in the chromosomes as a sequence of instructions. The standard GA operators are then applied to improve the code over time. In *Grammatical Evolution* (GE), the entire computer program is encoded in an integer string using a particular grammar: Each possible code entity is a certain integer value, and GE applies the GA operators to the string (chromosome). GE allows the optimization on the abstract genotype of the problem, while the standard GP works rather on the phenotype level on the direct code. The *Gene Expression Programming* (GEP) uses chromosomes of fixed length, which are coded as trees. Modified genetic operators avoid the creation of an invalid code. *Meta-genetic Programming* (MGP) gives more freedom to evolve how and where the genetic operators (recombination and mutation) can modify the chromosomes. MGP allows not only the optimization of the code, but also the optimization of the genetic optimization process by itself.

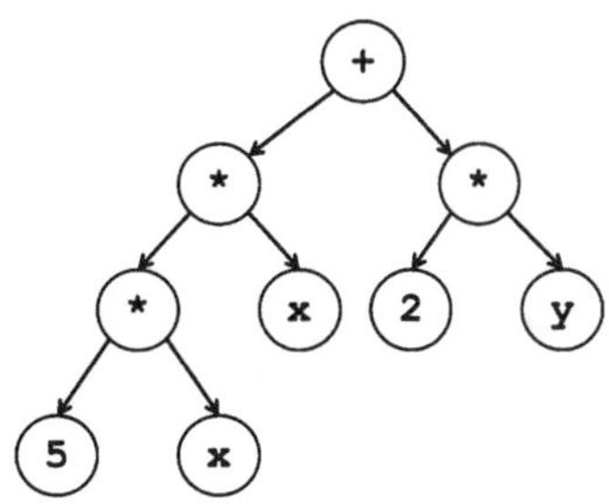

Fig. 2.5 Tree-structure representation of the function $5x^2 + 2y$

2.2.2.4 Evolutionary Programming

Evolutionary Programming (EP) was introduced in the 1960s by Lawrence J. Fogel, Alvin J. Owens, and M. J. Walsh (Fogel 1999). EP is a special kind of an ES, originally used to optimize the parameters of finite state machines for deriving computer programs or logic circuits. EP fully works on the level of phenotype and does not need a genotype representation. Thus, in contradiction to GP, EP aims to directly optimize the numerical parameters of a given problem (e.g. computer program), but not the structure of the source code itself.

EP mainly applies self-adaptive mutation to a set (i.e. the population) of specific individuals. In general, each parent individual is taken to generate a child individual. No classic crossover is applied, but a variation operator allows the combination of the information of different parents for the offspring. The selection is in reverse to the selection of EA: This EP selection, called the survivor selection, removes the individuals which will not to be taken into the next generation.

2.2.2.5 Multiobjective Evolutionary Algorithms

Non-dominated Sorting Genetic Algorithms (NSGA) and *Strength Pareto Evolutionary Algorithms* (SPEA) are variants of EA (respectively of GA) used in multiobjective optimization (also referred to as multicriteria or Pareto optimization), that solves optimization problems with more than one objective function to be optimized simultaneously. This kind of optimization is a kind of multiple criteria decision-making where the result of the optimization process is a multi-dimensional hyper-surface of non-dominated (Pareto optimal) and dominated solutions.

Two important example methods of EA, which are effective to solve such multiobjective optimization problems are the *Non-dominated Sorting Genetic Algorithms* (NSGA and the improved NSGA-II) (Deb et al. 2002) and the *Strength Pareto Evolutionary Algorithms* (SPEA and the improved SPEA2) (Zitzler et al. 2001). They use the advantage that a GA is a population-based method. Applied to multiobjective optimization problems, the individuals converge to the Pareto front, while the distribution mechanisms within these methods ensure that the single individuals (candidate solutions) are widely spread along the Pareto front.

2.2.2.6 Memetic Algorithms

Memetic Algorithms (Chen et al. 2011) (MA) are a set of hybrid algorithms combining EA (or at least any population-based approach) with local search or learning strategies, often deterministic ones. Examples of MAs are also mentioned as Baldwinian or Lamarckian EAs, cultural algorithms or genetic local search. The basic idea is that the evolution process of the EA is influenced, controlled or adapted by selectively modifying the individuals in the population within each iteration step. A meme is a cultural feature (e.g. a trait or behavior) passed on from person to person. Roughly speaking, MAs are extended EAs, where additional meme features provide additional strategies or knowledge about the problem in order to improve the optimization process.

The standard MAs (sometimes called MAs of the first generation) integrate an additional improved local search during the evolution process of the EAs, because EAs are effective in their population-based global search, but often not suited to converge exactly to the optimum. Consequently, near the optimum a local refinement strategy is more effective. Standard MAs combine these two strategies on the level of evolving the individuals (i.e. solution candidates) individually by selectively applying local search strategies.

The extended MAs, often referred to as multi-meme MAs (or MAs of the second generation) directly modify the individuals by applying possible meme features, i.e. additional knowledge about the problem, to the solution candidates. Within the iteration steps, not only the solution candidates improve, but also the kind and manner of memes which are applied. It is a kind of self-fulfilling prophecy: The individuals carrying the better meme features are superior, which leads to the point that these memes are selected to improve the solution candidates still further. While in this kind of MA the set of possible memes is predefined, self-generating MAs (or MAs of the third generation) develop themselves possible memes within the (meta-)optimization process.

2.2.2.7 Other Evolutionary Computation Techniques

In a *Learning Classifier System* (LCS) not only states or properties are evolved, but also rules. The rules themselves and the sequence of applying them are optimized to undertake certain actions in a particular environment. These rules are encoded in the chromosomes, which are modified using the GA operators. The principle of reinforcement learning computes the fitness value measuring the quality of these rules and the related actions in the particular environment.

Differential Evolution (DE) is related to EA, working with real-valued populations and operators to be used in ES and GA. DE is mainly used for optimization problems with real-valued multidimensional, often noisy and time-changing problems, where no gradient information is available. Its main extension lies in the generation of the child individuals (candidate solutions) using the information of three parent candidate solutions. The combination of information stored in these three parents allows guiding the optimization process faster in the direction of higher improvement.

Harmony Search (Geem et al. 2001) (HS) iteratively modifies the vector components in a set of solution candidates following the analogy of improvising melodious, harmonic music. Its specialty lies in the combination of all solution candidate vectors and the corresponding fitness to one matrix, the harmony memory. The entries of the matrix are selectively modified by applying methods of composing or arranging music, converging to a global optimum.

Other recent evolutionary computation techniques are *Neuroevolution* (evolving neural networks using principles of GP), *Evolutionary Robotics* (evolving controllers for autonomous robots), *and Learnable Evolution Model* (the evolution is guided by hypotheses rather than by the randomly applied EA operators for generating new individuals).

2.2.3 Swarm Intelligence

The interest in research in *Swarm Intelligence* (SI) started around the early 1990s (Meystell et al. 1990). The methods belonging to SI follow the principle of swarms, which is commonly stated as *optimization via emergence* or *the whole is more than its single components*. Emergence means here the appearance of patterns, larger entities, or regularities. Thereby, SI is copying the collective behavior of self-organized multi-agent systems, found in nature or natural processes: A population of simple agents (individuals, candidate solutions) are interacting locally, mostly following simple rules, and leading to an intelligent global behavior. The single individual has usually only a minor importance.

An intelligent behavior appears only after the interaction with other, mostly similar individuals. Stochastic mechanisms provide a variation within the population in order to explore the living environment (i.e. search space) and solution paths sufficiently without (almost) any additional knowledge about the problem. This swarm behavior is mostly used to solve optimization and search problems. A large applications area is logistics, where often combinatorial aspects lead to very large search spaces.

Swarm Algorithms (SA) are often mentioned as a sub-category of both, SI and EC, because they follow the principles of swarms and of evolution. Of late, SA are considered more and more as the representative method of SI. Sometimes, SA is even used as a synonym for SI.

The most important methods belonging to SI respectively SA are Particle Swarm Optimization (PSO), Ant Colony Optimization (ACO), Bees Algorithm (BA, recently more commonly denoted as Artificial Bee Colony (ABC) optimization) and Cuckoo Search (CS). Many other methods can, for instance, be found in the classification by Fister Jr et al. (2013). PSO knows further sub-methods such as binary, discrete or combinatorial PSO. The Firefly Algorithms (FA) and the very similar Glowworm Swarm Optimization (GSO) are considered as sub-methods of SA, too. Constantly, further sub-methods of SI respectively SA are emerging.

The application areas of SI methods mostly surround logistic or networking problems, such as routing, transportation and network configuration problems, but as well the simulation of crowds or heaps of particles and/or of particular items. Other application areas are the development of software controllers and the field of robots, consequently mentioned as swarm robotics.

2.2.3.1 Particle Swarm Optimization

Particle Swarm Optimization (PSO) is used to find the global optimum—in its standard form—in a continuous n-dimensional search space. In PSO, the candidate solutions are named *particles* instead of individuals as used in EC. These particles form a population, the swarm, which is moving through the search space. Iteratively, the particles of the swarm with the relatively better fitness attract the other particles to let the swarm move towards a local optimum while some particles search always the space for a better fitness.

In the basic variant of PSO, every particle p_i of the swarm of the size n is moving in the search space depending on its actual position x_i and its particular velocity vector v_i. The velocity vector of each particle, thus each individual's movement, is additionally influenced by the positions and velocities of the other particles in the swarm, preferably, the most by the recently best known particle(s) of the swarm. After an iteration, all the particles p_i move to a new position depending on their former position and their adapted velocity vectors. Another particle might now be the best particle of the swarm.

State of particle p_i with its initial position x_i and velocity v_i:

$$p_i = (x_i, v_i)$$

v_i^{new} is the adapted velocity of particle p_i depending on the velocity of the other particles j calculated with the influence factors a_{ij}. These influence factors mainly depend on the distance between the two particles and the relative fitness of particle j compared to particle p_i.

$$v_i^{new} = v_i + \sum_{j=1}^{n} a_{ij}\left(v_j - v_i\right)$$

New position x_i^{new} of particle p_i after the iteration step Δt is:

$$x_i^{new} = x_i + v_i^{new} \cdot \Delta t$$

An improvement of PSO was introduced by James Kennedy and Russell Eberhart (Kennedy and Eberhart 1995). They defined the new velocity v_i^{new} as a function of two summands added to the recent velocity. The first summand is history-related in the sense of the position p^{best}, where the particle p_i has found its best fitness in the search space. The second summand is defined as the position p^{global}, where the swarm has found the best fitness so far.

$$v_i^{new} = w_i v_i + c_i^{best} r_i^{best}\left(p_i^{best} - x_i\right) + c_i^{global} r_i^{global}\left(p^{global} - x_i\right)$$

Here, the coefficients c^{best} and c^{global} are acceleration factors which allow the particle to adapt more or less to the behavior of the entire swarm or to be more individual. The random coefficients r^{best} and r^{global} follow a uniform random distribution on the interval *[0, 1]*, while w is an inertia coefficient, normally reduced during the optimization process to increase the convergence ability.

2.2.3.2 Discrete Particle Swarm Optimization

The *Discrete Particle Swarm Optimization* (DPSO), also referred to as *Combinatorial PSO*, or in a digital case as *Binary PSO*, is extending PSO to discrete search spaces, as found in combinatorial problems. Here, the position x_i of the particle p_i is a solution vector of discrete numbers in an n-dimensional discrete search space. For example, in the combinatorial problem of arranging uniquely the five numbers {1, 2, 3, 4, 5} downward with respect to their value, a solution candidate might be $x_i = (3, 2, 4, 5, 1)$, where the optimal solution is $x_{opt} = (5, 4, 3, 2, 1)$.

In contrast to adding in each iteration a velocity-driven component to the recent position, the velocity vector v_i now defines one or even a set of exchange operations to be performed on the candidate solution x_i. E.g. $v_i = \{(1, 4)\}$ defines that the first and the forth entry are exchanged:

$$x_i^{new} = v_i(x_i) = v_i^{(1,4)}(x_i) = v_i^{(1,4)}(3, 2, 4, 5, 1) = (5, 2, 4, 3, 1)$$

2.2.3.3 Ant Colony Optimization

Ant Colony Optimization (ACO) was proposed by Marco Dorigo in the early 1990s (Colorni et al. 1991). In ACO, the individuals are named *ants*, which try to find the shortest way from their colony, the starting point, to a so-called food source through a graph of possible ways in the search space. Often, ACO is applied in scheduling, routing or assignment problems, where the sequence of discrete steps has to be optimized. Today, many different variants of ACO exist.

Originally, the foundations of the computational optimization algorithm ACO are highly inspired by the behavior of real ants: Ants mark their crawling trails using pheromones, an odorous substance, which is continuously set while the ants move on these trails. The pheromone level is slowly vaporizing over time. But, the more often the same ants or further new ants crawl along the same trail, the more intensive the pheromone level will be. Then, passing ants are influenced by the pheromone intensity and will—with higher probability—change their route to the more attractive trail with the higher pheromone level.

Every ant is representing a solution candidate vector x_i (the trail) as a sequence of discrete edges, e.g. $x_i = (A, D, B, C, E)$ from starting point A to endpoint E. At the beginning, all ants and pheromone levels on the edges are randomly initialized. The fitness of each ant is computed. Then, the ants (preferably only those with the better fitness values) add a new pheromone level to their trails, thus on all the edges of their trials. Afterwards, every ant is updating its solution candidate vector x_i, where some edges can be changed due to a trade-off between the part-fitness of the edge and the pheromone level set on it. From iteration to iteration, the ants tend to follow the sub-trails with stronger pheromone level. Over the time, the best route is found marked with the strongest pheromone level.

2.2.4 Neural Networks

The computational models, which simulate the functionality of real brains or at least of small parts of them on the computer, are called *Neural Networks* (NN). Literature sometimes uses the name *Artificial Neural Networks* (ANN) as well in order to distinguish between those neural networks of biological brains of animals and human beings, and the computational model of NNs. In this book, we use only the abbreviation NN for the computational models, omitting the "A" for the adjective "artificial".

NNs use the principle of natural cognition. Inspired by real biological neural networks of brains, NNs are processing information through a network of interconnected artificial neurons—in an abstract view similar to the information processing in biological brains. Various types of NNs exist, where the most prominent ones are Feedforward Neural Networks (FNN) respectively Perceptron Neural Networks (PNN), and Self-Organizing Maps (SOM).

The FNNs are the simplest NNs. They use a data flow from the neurons of the input layer through neurons of hidden layers (if there are any at all) to the neurons of the output layer. While multiplying the respective input values with the weights of the connections between the neurons, each neuron is summing up its input. An activation function defines a threshold when the neuron transmits an output value larger than zero. This simple algorithm is called perceptron. During a training phase of the NN prior to its application the weights and possibly other network parameters are adjusted. Using sets of training data all the weights in the NN are adjusted so that the input data is matched as good as possible to the wanted output data. The main aspect is back-propagation of some information of the transmitted data in order to train the NNs. In simple NNs the number of neurons per layer and the number of layers is predefined. High-level NNs allow to increase the number of neurons, layers and connections and thus the depth within the network.

SOMs are used for unsupervised mapping of high dimensional to low dimensional data, e.g. used in clustering problems. SOMs are often called Kohonen maps named after Teuvo Kohonen who introduced this kind of NNs in the 1980s (Kohonen 1982). In contrast to the classical NNs, SOMs use certain kinds of neighborhood functions to identify topological properties of the input vector. SOMs work with training data and then with mapping. The training data vectors build a map, which is then used to classify an unknown new data vector.

The main application areas of NNs are classification and optimization. Here, NNs implicitly evaluate the relationships between input and output data, finding patterns. NNs are fault- and noise-tolerant using non-linear threshold activation functions.

2.2.5 *Fuzzy Logic*

Fuzzy Logic (FL) applies the principle of fuzziness by assuming that logical values are not just true or false, or either 1 or 0, but can adopt any value from the interval [0, 1]. Similar to human cognition and human reasoning, FL allows a fuzzy modelling of the problem. FL uses a many-valued logic represented by a set of logical values allowing approximate values and incomplete or ambiguous data. FL is able to provide conclusions and approximate solutions based on fuzzy information. Its application areas are the modelling of fuzzy expressions, making inference based on fuzzy information, generating fuzzy sets of rules, and clustering of data. FL is often considered as fault- and noise-tolerant using weak rules.

2.2.6 *Artificial Immune System*

The methods which belong to *Artificial Immune System* (AIS) follow the principles of combatted pathogens in biological bodies. The human immune system identifies particular pathogens as a potential source of illness, which are then called antigens. Thus, AIS is inspired by the immune system of humans and animals. The AIS methods use a population of agents which learn to identify certain patterns of antigens that intrude a system. Over the generations, these agents improve via positive stimulated selection, clonal reproduction and adaptation to identify possibly changing pathogens as well. Therefore, the characteristics and processes of the immune system for learning and memory are transferred to AIS. Application areas of AIS are the development of adaptive systems and optimal behavior for problem solving as well as pattern recognition.

2.2.7 *Further Related Methods*

2.2.7.1 Reinforcement Learning

Reinforcement Learning (Kramer 2009) (RIL) follows the principle of "reward and punishment". Desired behavior is rewarded whereas undesired or not goal-oriented behavior is punished. Different learning strategies (e.g. dynamic programming, temporal difference learning) allow the system to iteratively learn a better behavior. Application areas are the control of characteristics and the learning of optimal behavior.

2.2.7.2 Simulated Annealing

Simulated Annealing is a generic probabilistic metaheuristic algorithm. Although this method is not inspired by the (living) nature, but by physics or more precisely by thermodynamics (the annealing process in metallurgy), we count Simulated Annealing to the CI methods. It follows the thermodynamic principle of a controlled cooling down of melted metal in order to increase the size of its arising crystals and reduce their defects. During the incremental steps of successively reducing the temperature, local minima or maxima are skipped in order to find the global optimum. The application areas of Simulated Annealing are in global optimization, where the search space is mostly large and discrete.

2.2.7.3 Further Metaheuristics: Local and Tabu Search

In the following, we briefly discuss further metaheuristics, which are not nature-inspired, but share similarities with several of those nature-inspired approaches portrayed above. An additional reason to include them in this book is that these metaheuristics belong to the most successful methods applied to various hard-to-solve problems in logistics.

Local Search, Neighborhood Search

One of the most common approaches in metaheuristics is the strategy of performing a *Local Search*—also denoted as *Neighborhood Search*. The underlying concept is that it is usually more promising to search improved solutions in the neighborhood of already determined solutions than in arbitrary regions of the entire search space.

Another reason is that local search is often used for large search spaces, which grow exponentially with the number of variables. While searching the whole set of feasible solutions is mostly not acceptable with respect to computation time, searching a small neighborhood area can be done very quickly. Obviously, local search requires the definition of neighborhoods of solutions, which depend on the particular problem type and can be referred to as distance between solutions or number of operations to transfer one solution into another one. Local search is—in its basic form—an iterative method, which is "jumping" from one solution to another better one in the neighborhood until a termination criterion (e.g. maximum number of iterations or inability to improve) is fulfilled. An obvious drawback of local search is that it can get stuck in a local, non-global optimum.

Quite often, local search is used as a background concept, on which other more complex metaheuristics are built, or which use local search as an embedded method for the improvement of intermediate solutions. For instance, simulated annealing uses local search in a more complex way and memetic algorithms include them in the framework of evolutionary algorithms. Also, typical mutations in evolutionary algorithms could be interpreted as terms of local search, because they usually prefer the determination of new solutions close to parent solutions.

Local search is further extended by combining it with other metaheuristics: In *Multistart Local Search* (Hart 1998), the problem of ending up in a local optimum, but not the global one, is tackled by restarting the local search from new (for instance randomly determined) starting solutions.

In *Iterated Local Search* (ILS) (Den Besten et al. 2001), the problem is treated by a particular perturbation of the found local optimum. The search continues repeatedly. In contrast to a multistart local search, the perturbed solutions should be sufficiently different from the previous solutions (to escape the local optima), but not "totally random".

Tabu Search

One of the most well-known advancements of local search is *Tabu Search* (Glover 1989, 1990), where the name "tabu" denotes forbidden. The central idea of this approach is to use a list of previously visited solutions, which are to be avoided in the subsequent search. Instead of storing visited solutions it is also possible to store some properties of previously visited solutions which are to be avoided in the subsequent search. The idea of a so-called tabu list addresses the problem that often, especially in combinatorial optimization problems, identical solutions are visited again and again during the local search. Apart from avoiding unnecessary computation time, the tabu list may also help to escape local optima together with the concept of allowing the search also to progress with worse solutions found during the iterations. (Nevertheless the best found solution is always stored for consideration at the end of the search process.) There are many variants with respect to tabu search, e.g. concerning the definition and length of the tabu list and its usage within the search.

Variable Neighborhood Search

Another successful and widely used variant of local search is the *Variable Neighborhood Search* (VNS) (Mladenović and Hansen 1997). VNS assumes that not only one, but several neighborhoods of different sizes exist. Usually, it is assumed that there are *k* neighborhoods, where a neighborhood with a higher index includes the neighborhoods with smaller indices.

VNS is based on the idea that local search should first take place in the smaller, closer neighborhood and then continue (when not being successful) with the next larger and wider neighborhood. Once a better solution is found the search continues, firstly again in its smallest neighborhood. This helps avoid getting stuck in a local optimum of a very small neighborhood, but also avoids spending too much computation time for search in large neighborhoods.

There are several variants of VNS. For instance, in *Basic VNS* a single, randomly determined point from a considered neighborhood is improved by local search before probing the next neighborhood. In *Reduced VNS* the embedded step of local search is omitted, i.e. if a randomly selected point is not directly better, the search continues with the next neighborhood. Variable Neighborhood Descent (VND) is based on the idea of doing a deterministic local search in a considered neighborhood.

Greedy Randomized Adaptive Search Procedure (GRASP)

As a last example of frequently used metaheuristics let us mention the *Greedy Randomized Adaptive Search Procedure* (GRASP) (Feo and Resende 1995), which is based on iterations for constructing and then improving solutions by local search. The construction phase is problem specific. The general idea is given by adding "solution elements" successively, which increase their "quality" in a greedy way. While adding these solutions elements, a particular randomness is included in order to maintain variability.

References

Büche, D., Stoll, P., Dornberger, R., & Koumoutsakos, P. (2002). Multiobjective evolutionary algorithm for optimization of noisy combustion processes. *IEEE Transactions on Systems, Man, and Cybernetics Part C: Applications and Reviews, 32*(4), 460–473.

Burke, E. K., & Kendall, G. (Eds.). (2006). *Search methodologies – introductory tutorials in optimization and decision support techniques*. New York: Springer.

Chen, X., Ong, Y.-S., Lim, M.-H., & Tan, K. C. (2011). A multi-facet survey on memetic computation. *IEEE Transactions on Evolutionary Computation, 15*(5), 591–607.

Colorni, A., Dorigo, M., & Maniezzo, V. (1991). Distributed optimization by ant colonies. In *Proceedings of the First European Conference on Artificial Life* (Vol. 142, pp. 134–142). Paris, France: Elsevier Publishing.

Deb, K., Pratap, A., Agarwal, S., & Meyarivan, T. (2002). A fast and elitist multiobjective genetic algorithm: NSGA-II. *IEEE Transactions on Evolutionary Computation, 6*(2), 182–197.

Den Besten, M., Stützle, T., & Dorigo, M. (2001). Design of iterated local search algorithms. In *Applications of evolutionary computing* (pp. 441–451). Berlin/Heidelberg: Springer.

Dornberger, R., Büche, D., & Stoll, P. (2000). Multidisciplinary optimization in turbomachinery design. In *European Congress on Computational Methods in Applied Sciences and Engineering (ECCOMAS 2000)*. Barcelona, Spain.

Engelbrecht, A. P. (2007). *Computational intelligence: An introduction*. Chichester: Wiley.

Feo, T. A., & Resende, M. G. (1995). Greedy randomized adaptive search procedures. *Journal of Global Optimization, 6*(2), 109–133.

Fister, I., Jr., Yang, X. S., Fister, I., Brest, J., & Fister, D. (2013). A brief review of nature-inspired algorithms for optimization. *Elektrotehniski Vestnik/Electrotechnical Review, 80*(3), 116–122. arXiv preprint arXiv:1307.4186.

Fogel, L. J. (1999). *Intelligence through simulated evolution: Forty years of evolutionary programming*. New York: Wiley Series on Intelligent Systems.

Fulcher, J., & Jain, L. C. (Eds.). (2008). *Computational intelligence: A compendium* (Studies in computational intelligence, Vol. 115). Berlin/Heidelberg: Springer.

Geem, Z. W., Kim, J. H., & Loganathan, G. V. (2001). A new heuristic optimization algorithm: Harmony search. *Simulation, 76*(2), 60–68.

Glover, F. (1989). Tabu search – part I. *ORSA Journal on Computing, 1*(3), 190–206.

Glover, F. (1990). Tabu search – part II. *ORSA Journal on Computing, 2*(1), 4–32.

Goldberg, D. E. (1989). *Genetic algorithms in search, optimization, and machine learning*. Boston: Addison-Wesley.

Hansen, N., & Ostermeier, A. (2001). Completely derandomized self-adaptation in evolution strategies. *Evolutionary Computation, 9*(2), 159–195.

Hart, W. E. (1998). Sequential stopping rules for random optimization methods with applications to multistart local search. *SIAM Journal on Optimization, 9*(1), 270–290.

Holland, J. H. (1975). *Adaptation in natural and artificial systems: An introductory analysis with applications to biology, control, and artificial intelligence*. Ann Arbor: University of Michigan Press.

Meystell, A., Herath, J., & Gray, S. (Eds.). (1990). *Proceedings of 5th IEEE International Symposium on Intelligent Control 1990*. New York: IEEE.

Kennedy, J., & Eberhart, R. (1995). Particle swarm optimization. In *IEEE International Conference on Neural Networks, 1995* (Vol. 4, pp. 1942–1948). New York: IEEE.

Kohonen, T. (1982). Self-organized formation of topologically correct feature maps. *Biological Cybernetics, 43*(1), 59–69.

Koza, J. R. (1992). *Genetic programming: On the programming of computers by means of natural selection*. Cambridge, MA: MIT Press.

Kramer, O. (2009). *Computational intelligence – Eine Einführung*. Berlin/Heidelberg: Springer.

Kruse, R., Borgelt, C., Klawonn, F., Moewes, C., Steinbrecher, M., & Held, P. (2013). *Computational intelligence – a methodological introduction*. London: Springer.

McCarthy, J. (2007). *What is artificial intelligence*?. Accessed April 17, 2014, from http://www-formal.stanford.edu/jmc/whatisai/whatisai.html

McCarthy, J., Minsky, M., Rochester, N., & Shannon, C. (1955). *A proposal for the Dartmouth summer research project on artificial intelligence*. Accessed November 30, 2013, Archived from the original on http://www-formal.stanford.edu/jmc/history/dartmouth/dartmouth.html

Mitchell, T. M. (1997). *Machine learning*. New York: McGraw-Hill.

Mladenović, N., & Hansen, P. (1997). Variable neighborhood search. *Computers & Operations Research, 24*(11), 1097–1100.

Neapolitan, R. E., & Jiang, X. (2012). *Contemporary artificial intelligence*. Boca Raton, FL: Chapman and Hall.

Poole, D. L., Mackworth, A. K., & Goebel, R. (1998). *Computational intelligence: A logical approach*. Oxford: Oxford University Press.

Rechenberg, I. (1973). *Evolutionsstrategie – Optimierung technischer Systeme nach Prinzipien der biologischen Evolution*. Stuttgart: Frommann-Holzboog.

Rechenberg, I. (1994). *Evolutionsstrategie '94*. Stuttgart: Frommann-Holzboog.

Rozenberg, G., Bäck, T., & Kok, J. N. (Eds.). (2012). *Handbook of natural computing*. Berlin/ Heidelberg: Springer.

Schwefel, H.-P. (1977). *Numerical optimization of computer models*. Basel: Birkhäuser Verlag.

Zitzler, E., Laumanns, M., &Thiele, L. (2001). *SPEA2: Improving the Strength Pareto Evolutionary Algorithm*. TIK-Report 103. Zurich: Computer Engineering and Networks Laboratory (TIK), Swiss Federal Institute of Technology (ETH).

Chapter 3
Transportation Problems

Abstract The field of transportation belongs to the most important areas of logistics. Transportation problems occur in various variations and complexity and require a careful planning due to their significance and the need to solve them efficiently, i.e. with high quality and low costs. In general, transportation planning requires to determine the kind and quantity of the goods that are shipped, from where to where, at which time and by using which resources, in particular by which vehicles. The determination of a specific path or route belongs to the field of transportation planning as well.

In many cases some of these aspects are determined a priori, for instance, the origin and the destination are frequently given. We also assume that decisions on production and inventory are done before and not considered in combination with the planning of transports. (Some of such combined planning problems are considered in later chapters of the book.)

Unfortunately, many variants of transportation problems are hard from a computational point of view. This means that there are no algorithms that are able to find an optimal solution with an acceptable amount of running time, at least not for larger instances of transportation problems.

In the following we will take a closer look at some major variants of transportation problems and at methods that are able to solve them, at least with a sufficient quality. First we discuss assignment problems which allow a transport planning on a rather rough level mapping origins of goods and destinations based on transport cost. Next the comparably simple, but widely used shortest path planning tasks are discussed. In the next sections, the most important types of more complex tour planning problems are discussed: the travelling salesman problem and vehicle routing problems. In particular for the vehicle routing problem several practically relevant versions are considered including pickup and delivery problems. The chapter ends with a brief discussion of network flow problems.

T. Hanne, R. Dornberger, *Computational Intelligence in Logistics and Supply Chain Management*, International Series in Operations Research & Management Science 244,
DOI 10.1007/978-3-319-40722-7_3

3.1 Assignment Problems

One of the earliest problems considered in the field of transportation planning is to determine from which origin to which destination goods should be shipped assuming that a demand for goods is given in specific locations and that the required goods are available in other locations. It is also assumed that the distances between the different locations are specified, i.e. there is no need for a more detailed specification of transportation paths. The usage of specific vehicles or types of vehicle is not taken into account either.

The goal of this type of problem is only to determine which quantity of a given good should be shipped from which locations (with goods being available) to which other locations where there is demand for that good in order to satisfy the demand and to minimize the transportation costs. Usually, it is assumed that a sufficient total quantity of goods is available in order to satisfy all demand.

Let us consider a simple version of this type of a transport-related assignment problem which is also denoted as Linear Sum Assignment Problem (LSAP) (Burkard and Cela 1999). Suppose that we have n depot locations of goods, denoted as set L^D, and n consumer locations, denoted as set L^C, who require the goods from one of the depots. We assume that a cost function is given which measures the cost of a single shipment from a depot i to a consumer j, i.e.,

$$c_{ij} := c\left(l_i^d, l_j^c\right) \text{ for } i \in \{1, \ldots, n\}, j \in \{1, \ldots, n\}, l_i^D \in L^D, l_j^C \in L^C. \tag{3.1}$$

For instance, the cost could be measured by the Euclidean distance between the locations of l_i^D and l_j^C.

There are some more assumptions which, in the general case, do not appear to be particularly realistic: We assume that each consumer requires one single shipment from one depot and that each depot can deliver exactly to one customer. Moreover, shipments cannot be split. Apart from that there are no other restrictions to be observed.

The transportation problem is then to find suitable assignments of depots to customers which minimize the total cost function. The assignment can be expressed by variables x_{ij}, $i \in \{1, \ldots, n\}, j \in \{1, \ldots, n\}$, with $x_{ij} = 1$ if customer j is supplied by depot i, and $x_{ij} = 0$ otherwise. All variables can be briefly notated by an $n \times n$ matrix X.

The total cost function to be minimized can then be specified by

$$\min c(X) = \sum_{i=1}^{n} \sum_{j=1}^{n} c_{ij} x_{ij} \tag{3.2}$$

with the following restrictions for the variables:

$$\sum_{i=1}^{n} x_{ij} = 1, \text{ for all } j \in \{1, \ldots, n\}, \tag{3.3}$$

$$\sum_{j=1}^{n} x_{ij} = 1, \text{ for all i} \in \{1, \ldots, n\}, \tag{3.4}$$

$$\text{with } x_{ij} \in \{0, 1\} \text{ for all } i, j \in \{1, \ldots, n\}. \tag{3.5}$$

This simple assignment problem (LSAP) can be solved by various methods. Linear programming methods belong to these approaches as the problem can be relaxed to a linear problem by replacing (3.5) by $x_{ij} \geq 0$ for all $i,j \in \{1, \ldots, n\}$. Thus, the LSAP can be solved by available linear programming approaches such as the simplex algorithm and variants. Although these methods are not polynomial in the general case, the LSAP can be solved in polynomial time due to the utilization of special properties of this problem type. Among these polynomial time methods, the best known one is the so-called Hungarian method which requires a running time of $O(n^4)$. This method works by successive row and column reductions which allow to determine which variables are set to one and which are set to zero. There are improvements of this method with a smaller running time of $O(n^3)$.

3.2 Shortest Paths

Finding a shortest path is one of the most common problems in transportation and related areas. Given two locations A (starting point) and B (destination) which are connected by a network (or a graph) the task is to find a connected sequence of roads (or edges in general) with a shortest distance. The distance is defined by the sum of the lengths of the connected road segments (or, respectively, the edges of the graph). Instead of minimizing the distance, one can also minimize the time or costs of the used roads of the network.

Let $G = (V, E, c)$ be a weighted graph, that is: V is a set of nodes (also denoted as vertices) i.e. the crossing points of roads, $E \subseteq V \times V$ is a set of edges between the nodes (i.e. road segments which connect nodes) and c, c: $E \rightarrow R$ is a cost function for the edges (e.g. the distance between two directly connected nodes or the time needed for travelling the distance between the nodes or any other kind of cost measure). Usually, it is assumed that the costs are nonnegative, i.e. $c(e) \geq 0$ for all $e \in E$.

The task is then to find a sequence of edges, $e_1, e_2, \ldots, e_n$ with $e_1 = (A, v_2)$, $e_2 = (v_2, v_3), \ldots, e_{n-1} = (v_{n-1}, v_n)$, $e_n = (v_n, B)$ (i.e. the sequence starts in the starting location A and ends in the destination point B), and each edge ends in the starting point of the successive edge such that the costs are minimized.

Fortunately, such kind of problems can be solved with a reasonable amount of time. One of the best known algorithms to solve a shortest path problem is the Dijkstra's algorithm. This method works as follows:

Step 1: (Initialization)

a. Every node is assigned two labels, one for the *tentative distance* initialized by infinity and a Boolean label "*visited*" initialized as false, for node A the *tentative distance* label is initialized by 0 and the *visited* label is set to true.
b. The set "*unvisited*" is initialized by all nodes with visited = false.
c. The variable "current node" is initialized by A.

Step 2:

For the "current node" it is checked for all unvisited neighbors whether the distance of the current node + the distance from the current node to the considered node is smaller than the *tentative distance* of the considered node. In that case the *tentative distance* of the considered neighbor node is updated by this smaller value.

Step 3:

For the current node *visited* is set to true and the node is removed from set *unvisited.*

Step 4:

The algorithm stops when the current node was B or if the smallest distance among the nodes in the *unvisited* set is infinity (in that case there is no connection between A and B).

Step 5:

Select from the set *unvisited* a node that is marked with the smallest *tentative distance* and set it as the "current node".

Step 6:

Go to step 2.

Note that the algorithm is often used to determine the smallest distance of A to all other nodes and that the label "*tentative distance*" shows the smallest distance when the respective nodes are removed from the set *unvisited.* In the general case, the run time of the algorithm can be considered as quadratic with respect to the number of nodes if the costs are nonnegative, i.e. being in $O(n^2)$ if n is the number of nodes. Tighter approximations are possible when considering a specific data structure or taking into account both the number of edges and the number of nodes. [For the run time estimate $O(n^2)$ it is taken into account that the number of edges is maximally n^2.] For the calculation of all shortest paths (for all pairs of nodes) of a graph the runtime can be reduced to $O(n^2 \log n)$ using appropriate data structures (Takaoka 2013).

Shortest path problems are today implemented in various software tools. In particular, they are the backbone for today's GPS-based navigation systems. Such tools have an actual road network represented in their data and allow locating the current position of a vehicle or a person. When entering a destination, the software can calculate the shortest or fastest way to the destination making some assumptions of the speed of the vehicle (or the person). Also the specific kinds of roads (e.g. small, slow roads or faster highways), information about the different speeds on the roads according to the direction of travel (or not allowing some direction for one-way roads) and, in some case, information about the current traffic load are taken into consideration.

One issue of such kind of problems is the considerable size of real road networks. Although a shortest path algorithm is quite fast, the problem would be too time consuming when considering a real road network, at least when A and B are quite distant. Since a user assumes a quick calculation of a shortest path, a software embedded in navigation tools makes use of further refinements such as a simplification of the road network in order to provide an almost real time answer for a routing request. Also for the shortest path algorithm adaptations are made frequently, in particular, often these algorithms are rather based on A* heuristics which utilize information about a most promising exploration of the network. Other heuristic approaches used for speed-up are, for instance, branch pruning and the decomposition of the search problem (Fu et al. 2006).

3.3 The Travelling Salesman Problem

The travelling salesman problem is a very well-known problem in the field of transportation planning. However, many problems from other areas can be expressed as travelling salesman problems as well. Moreover, the problem is quite simple to describe but nevertheless computationally hard (NP-hard), such that it has become one of the most favourite problems for researchers trying to find good heuristics or approximations for hard problems. However, for the TSP in its general formulation it is assumed that even approximation algorithms with known quality do not exist (Lenstra and Kan 1981).

The problem can be described as follows: During a single trip a number of locations are to be visited and a shortest sequence of these locations is to be determined. It is assumed that the trip starts and ends in a given location called depot, and that no location needs to be visited twice. The distances between the different locations including the starting point are assumed to be given. Very often these distances are expressed in the form of a (square) matrix with rows and columns corresponding to the considered locations. Another possibility is that these distances are calculated, e.g. according to a metric like the Euclidean metrics and using given coordinates of the locations.

In any case, it is assumed that a set of locations $L = \{l_0, l_1, \ldots, l_n\}$ is given together with the distances (or the relating costs), i.e. a real-valued distance function $d{:}L \times L \rightarrow \mathfrak{R}$. Moreover, it is usually assumed that the distances are nonnegative, i.e. $d(a,b) \geq 0$ for all $a, b \in L$ and that the triangle inequality holds, i.e. $d(a,b) + d(b,c) \geq d(a,c)$ for all $a, b, c \in L$. In this notation, l_0 is typically used to indicate the depot, i.e. the starting location and the end location of the tour. Since

Table 3.1 Example distance matrix

	A	B	C	D
A		12	35	25
B	12		25	10
C	35	25		20
D	25	10	20	

starting and end locations are fixed, it is usually sufficient to specify a tour just by the sequence of locations visited in between which can be expressed by a permutation π of the index numbers $\{1, \ldots, n\}$ which represent the sequence $[\pi(1), \ldots, \pi(n)]$.

Example:

A supplier wishes to ship goods to three customers (locations B, C and D) from the warehouse (location A). The total shipments fit into the used vehicle. The distances (in minutes) are specified by the following matrix of travel times (Table 3.1):

The goal is now to determine the sequence in which the customers should be served in order to minimize the total travel time.

Before we consider the solution of the problem let us note that the above matrix is symmetric, that means a matrix coefficient in row i and column j is always equal to the coefficient in row j and column i. That means, going from one location to another location always takes the same time as going into the opposite direction. Of course, in reality this is often not the case. For instance, a different traffic load depending on the direction of travel, problems due to construction sites which only affect one direction or one-way roads can lead to different times according to the chosen direction between two points. Of course, an asymmetric formulation of the TSP can be considered as well.

When considering the solution of the TSP we basically see that there are $n!$ sequences for visiting n customers which are all feasible but usually different with respect to the objective function. If symmetry is assumed (as in the matrix above) the number of really different solution is reduced to $n!/2$. (Taking a tour according to any sequence is as long as travelling the inverse sequence.)

In our example there are six possible solutions which are as follows:

1. ABCDA: 78' (=12 + 25 + 20 + 21)
2. ABDCA: 77' (=12 + 10 + 20 + 35)
3. ACBDA: 95' (=35 + 25 + 10 + 25)
4. ACDBA: 77' (corresponds to solution 2 due to symmetry)
5. ADBCA: 95' (corresponds to solution 3)
6. ADCBA: 78' (corresponds to solution 1)

Obviously, solutions 2 and 4 are the best.

The distance between two locations which are given in the matrix can be a broad approximation but may also be based on solving a shortest path problem a priori, i.e. the shortest paths between all considered locations can be calculated first.

As the name Travelling Salesman suggests a typical scenario for such a problem could be the planning of a route for a salesperson who wishes to visit several customers during a tour. It is assumed that the customers can be visited in an arbitrary order so that it makes sense to determine an order with a smallest travelling time, travelling costs, or with the shortest total distance.

Another scenario might be the delivery of goods from one location (e.g. a production plant) to several customers (e.g. shops) or the collection of goods (e.g. letters or parcels) from a number of locations. As the problem does not assume a capacity limit the usage of a TSP formulation is, however, only suitable if capacity does not matter. Thus, it is assumed that everything that has to be transported always fits into the vehicle.

Other restrictions which may complicate the TSP are, for instance, time windows to be observed for vising the customers or the additional consideration of handling times at stops (e.g. for unloading shipments from the vehicle).

Let us consider the above example while assuming that 10 min are required for each unloading of a shipment. The optimality of the above solutions is, however, not affected by this. Every tour just takes 30 min longer due to three planned stops including unloading time. But if we assume time windows the situation may change: Let us assume that a tour starts at 8:00 at the warehouse and that D requires the goods to arrive between 8:15 and 8:45 while C cannot accept the shipment before 9:00.

Our two optimal solutions from above would be scheduled as follows:

1. Optimal solution:

$$A_{\text{departure}}\ 8:00\quad B_{\text{arrival}}8:12\quad B_{\text{departure}}8:22\quad D_{\text{arrival}}8:32\quad D_{\text{departure}}\ 8:42$$
$$C_{\text{arrival}}9:02\quad C_{\text{departure}}9:12\quad A_{\text{arrival}}9:47$$

2. Optimal solution:

$$A_{\text{departure}}\ 8:00\quad C_{\text{arrival}}8:35\quad C_{\text{departure}}8:45\quad D_{\text{arrival}}9:05\quad D_{\text{departure}}\ 9:15$$
$$B_{\text{arrival}}9:25\quad B_{\text{departure}}9:35\quad A_{\text{arrival}}9:47$$

While the first optimal solution can be realized (without any changes) the second previously optimal solution does not fulfill the time window constraints.

Let us note at the end of this section that there are various variants of the standard TSP. For instance, we could assume the locations to be visited by one of multiple travelling salesmen or vehicles and, moreover, we could assume that these operate from different depots (multiple depot, multiple travelling salesmen problem; see, e.g., Yadlapalli et al. 2009). Then there are TSP variants which consider pickups and deliveries of goods (see, e.g., Gendreau et al. 1999) or which do not require visiting all locations (Feillet et al. 2005). Travelling salesman problems could be formulated using different or multiple objectives (see, e.g., Shim et al. 2011) and some of the information from the TSP could be modelled as stochastic or fuzzy. For instance, a fuzzy distance matrix is used in Crisan, and Nechita (2008). For some part, these extensions and variants of the TSP serve requirements from practical problems. We will consider some of them in more details in Sect. 3.5 where we treat vehicle routing problems which are in many cases closer to practical applications than the TSP.

3.4 Methods for Solving the Travelling Salesman Problem

As mentioned before, the TSP belongs to the class of NP-hard problems. That means there is no known algorithm which can solve the problem to optimality within a time polynomially dependent on the problem size (and, moreover, such an algorithm does probably not exist). That means, for problems of a size not too small, it is not possible to solve the problem to optimality with a limited amount of time and even approximations with defined quality remain difficult.

For that reason, heuristics and metaheuristics play a major role in treating the problem and many respective algorithms have been developed and explored for the TSP during the last decades. Although such algorithms do not allow for a proven best solution (at least not with subexponential runtime), the obtained solution is usually good enough for practical purposes. In fact, even for larger problem sizes, it could be shown in many cases that optimal or almost optimal solutions could be achieved. Very often, the algorithms were tested using a widely known library of test problems available called TSPLIB—A Travelling Salesman Problem Library (cf. Reinelt 1991) which makes the comparison of different algorithms easier.

As for many other hard optimization problems, there are three basic classes of methods being used for solving the TSP: exact (mathematical) methods, metaheuristics and other CI approaches, and simple heuristics. Among the exact approaches, branch and bound, branch and cut and similar approaches can be found. Simple heuristics are for instance based on route improvement operations which can also be integrated into more complex metaheuristics. Among the metaheuristics, basically the full range of methods belonging to that class has been explored for the TSP.

3.4.1 Heuristics for the Travelling Salesman Problem

Before we consider further details of such methods, let us discuss some principles of more simple heuristics. First of all, there are construction heuristics such as the "nearest neighbor" approach: Starting from the depot, the nearest customer location is visited first. From there, the nearest of the remaining (unvisited) locations is visited and so on until all locations are planned for the tour.

Secondly, there are improvement heuristics. The most simple of them is the exchange heuristics denoted as 2-opt. Let us assume that there are two positions in a tour, $p, q \in \{1,\ldots,n\}$, with $p < q$, and that the following condition for their distances or travel times is fulfilled: $d(l_{p-1}, l_p) + d(l_q, l_{q+1}) > d(l_{p-1}, l_q) + d(l_p, l_{q+1})$, then reordering the tour by flipping the partial sequence $(l_p, \ldots, l_q)$ to the new tour $(l_0, \ldots, l_{p-1}, l_q, l_{q-1}, \ldots, l_{p+1}, l_p, l_{q+1}, \ldots, l_{n-1})$ reduces the total travel times (or costs). The respective heuristics searches repetitively for such situations in the sequence of locations and modifies the sequence accordingly until a local optimum is reached. In a similar way, a 3-opt heuristics with the exchange of three edges can be defined.

The Lin-Kernighan heuristic tries to build a sequence of such single flip operations in order to improve the tour. The main difference is that it is not required for each single flip operation to improve the tour. Only after performing the whole sequence, the resulting tour should be better than the original sequence. The creation of such sequences can, for instance, be done with a backtracking approach (Applegate et al. 2003). Of course, it is reasonable to limit the backtracking depth or the maximum length of sequences to avoid an almost complete exploration of the search space. As shown by Applegate et al. (2003) using sequences of up to 25 single flip operations leads to very good results even for large TSP instances. Together with some further improvements the approach is usually denoted as chained Lin-Kernighan heuristics.

3.4.2 *Evolutionary Algorithms for the Travelling Salesman Problem*

Very often, metaheuristics applied to the TSP (and other tour building problems) make use of more such simple heuristics. Let us discuss this for the example of evolutionary algorithms which played a major role in studies on the TSP and which have been used in many variants with the earliest publications dating back to the 1980s (see, e.g., Grefenstette et al. 1985; Braun 1991; Larrañaga et al. 1999; Tsai et al. 2004; Snyder and Daskin 2006).

Relevant design principles of evolutionary algorithms (and other metaheuristics) are, in particular, the random variation operators mutation and crossover. If these operators are not based on an appropriate encoding of solutions and/or taking care of further adaptation of their standard forms, they would usually lead to infeasible solutions (see Chap. 5 on Scheduling for further examples) which would increase the computational burden and decrease the efficiency of the evolutionary method. Therefore, it is useful to construct such operators which automatically produce feasible solutions, i.e. tours with each node occurring exactly once in the tour.

One possibility to avoid such requirements of designing suitable variation operators which maintain feasibility (but make the respective algorithms more complex and require additional computational time) is to use a more suitable problem representation, e.g., the random key encoding (Snyder and Daskin 2006, also cf. Chap. 5 on Scheduling). Here, every location is associated with a real number (floating-point number) which determines its position in a respective sequence. That means, for determining the sequence corresponding to a solution the locations need to be sorted according to their associated numbers. Variations of sequences are derived from variations of the assigned numbers and the usual mutation and crossover operators can be applied. For instance, the number can be subject to normally distributed mutations (e.g. as in the concept of evolution strategies), i.e. using normally distributed random variables that are added to the respective real-values denoting the later position of locations in a sequence. Crossovers take place only for the array of associated numbers, e.g. in form of one-point crossovers or two-point crossovers. In any case, the result of such a variation is always again a sequence without double or missing locations. Of course, if further constraints are to be considered (e.g. capacity constraints) this must be taken into account additionally.

Among the disadvantages of using this representation, the additional effort for sorting a sequence (especially when it is long) and possible strong effects of the variation operators are occasionally mentioned (Chatterjee et al. 1996). With respect to the second disadvantage we can note that this effect can be effectively controlled when using $N(0,\sigma)$-distributed mutations and, in particular, small σ values should only lead to smaller changes in a sequence.

In most studies on using evolutionary algorithms for the TSP (and similar problems), however, a "direct" representation of solutions in form of sequences is chosen and evolutionary operators are adapted in a suitable manner:

Simple variation operators such as mutations in evolutionary algorithms must take care that a possibly random change does not lead to an unwanted double consideration of the same location or to forgetting a location. When a problem solution is represented as a sequence of locations (or their respective indices) at least two positions need to be changed during a mutation, e.g. an exchange of two randomly selected nodes in a tour (exchange mutation).

Further frequently used possibilities of mutation operators are inversions, displacement mutations, insertion mutations, and scramble mutations (see, e.g. Larrañaga et al. 1999; Ahmed 2010). In the inversion mutation (sometimes also denoted as simple inversion mutation), a tour is reversed between two randomly selected positions, e.g. if for a sequence (without depot)

$$(2,\ 6,\ 5,\ 4,\ 3,\ 1)$$

the positions 3 and 5 are determined, the following sequence results:

$$(2,\ 6,\ 3,\ 4,\ 5,\ 1).$$

The insertion mutation operator selects a random position in a considered sequence, removes it from the sequence and re-inserts it at a new randomly determined position. This idea is generalized by the displacement mutation operator (also denoted as cut mutation) which selects a random subsequence from a solution and shifts it to a new randomly determined position in the tour. Another type of mutation (sometimes also confusingly denoted as inversion mutation) combines the displacement and simple inversion mutation, i.e. a randomly selected subsequence is removed and reinserted at a new random position but in reverse order. Finally, let us mention the scramble mutation operator which determines a random subsequence and randomly reorders this subsequence.

With respect to crossover, the standard operators cannot be used for sequence-based representations: Splitting two sequences at the same position (one-point crossover), for instance, and connecting the different parts of the two solutions (one-point crossover) usually leads to an infeasible solution due to double or missing locations in the resulting sequences. Therefore, specific crossover operators which ensure feasibility are required. A survey of different suitable recombination and crossover operators is given in Larrañaga et al. (1999): One often used type is the partially mapped crossover (PMX). In a respective crossover two cut points are randomly selected for the two parent solutions to be recombined. In the offspring, the respective sequences between the cut points are exchanged for the two offspring solutions. For the other positions in a tour the original locations of the respective parents are used unless these locations are already in the exchanged sequence. In this case, the respective values are replaced by the original values from the positions in the exchange sequence. Let us consider an example:

If for the sequences (2, 6, 5, 4, 3, 1) and (1, 2, 6, 4, 3, 5) the cut points between the first and the second and between the third and the fourth position are selected, for the offspring solution first the intermediate ranges with positions two and three are set from the "opposite" parent, i.e. (*, 2, 6, *, *, *) and (*, 6, 5, *, *, *). Then for the remaining positions of first offspring the respective values from the first parent are copied. At position one we find that "2" already occurs in the exchange sequence (at position 2), and therefore the original value from that position should be used, i.e. "6". As this value is also in the exchanged sequence at position 3, we use the original value from that position, i.e., "5". The values from positions 4–6 can be used as they do not appear in the exchanged sequence. For the second offspring, all original positions except position 6 can be used. Here similar considerations as before lead to a placement of "2" at this position. That means, the offspring solutions result as (5, 2, 6, 4, 3, 1) and (1, 6, 5, 4, 3, 2).

The order crossover (OX1) operator works initially similar: For two parent solutions two cut points are determined randomly. The cut points for the two parents can be different but they should include a sequence of identical length. Then the sequences between the cut points are exchanged for the respective offspring solutions. After that, starting after the second cut point the remaining positions are filled step-by-step with the next locations in the original sequences, leaving out locations which already occur in the exchanged sequences.

For instance, if for the sequences (2, 6, 5, 4, 3, 1) and (1, 2, 6, 4, 3, 5) the cut points after positions one and three (first parent), and after position three and five (second parent) are selected, then for the offspring solution first these positions are set, i.e. (*, 4, 3, *, *, *) and (*, *, *, 6, 5, *). Starting with the first offspring at position 4, the original values at this and the next position, i.e. "4" and "3" cannot be used as they appear in the exchanged part. Therefore "1" is chosen, then "2", "6", and "5", and the resulting sequence is (5, 4, 3, 1, 2, 6). Note that at the end of the sequence the process continues at the beginning of the sequence until the first cut position is reached and all positions are set. For the second offspring the result is as follows: (2, 4, 3, 6, 5, 1).

A similar idea of order conservation is used in the order based crossover (OX2). Here some locations in one parent solution are determined randomly. For the offspring of the other parent solution, all positions are copied except the positions corresponding to the locations determined from the other parent solution. Those positions are now set by these locations but in the sequence coming from the other parent.

Let us assume that for our example the sequences (2, 6, 5, 4, 3, 1) and (1, 2, 6, 4, 3, 5) the positions 1, 3, and 5 are determined from the second parent. These correspond to the locations "1", "6", and "3". Thus for the offspring of the first parent, we first only set the remaining locations, i.e. (2, *, 5, 4, *, *). The locations "1", "6", and "3", are set in the sequence from the second parent and we obtain the sequence (2, 1, 5, 4, 6, 3). When considering the same random positions for the first parent, the second offspring solution leads to the following offspring result: (1, 2, 6, 4, 5, 3).

In the position-based crossover (POS) also a random set of positions from the parent solutions are determined first. For the first offspring, the locations at these positions are copied from the second parent while—starting at the beginning of the sequence—the remaining positions are filled step-by-step by locations according to the sequence of the first parent taking care that no location is chosen twice.

If in our example sequences (2, 6, 5, 4, 3, 1) and (1, 2, 6, 4, 3, 5) again the positions 1, 3, and 5 are randomly determined, we first obtain for the first offspring the sequence (1, *, 6, *, 3, *) which is filled up to (1, 2, 6, 5, 3, 4). For the second offspring we obtain first (2, *, 5, *, 3, *) and then (2, 1, 5, 6, 3, 4).

Among the various other crossover operators found in the literature (see, e.g., Larrañaga et al. 1999) are, for instance, the cycle crossover operator (CX), heuristic crossover operators (which use probability distributions based on edge costs), the genetic edge recombination crossover (ER), the edge recombination crossover (ERX), the sorted match crossover operator, the maximal preservative crossover (MPX), the voting recombination crossover (VR), the alternating position crossover (AP), and the generalized N-point crossover (GNX).

Respective mutation and crossover operators can be significantly more complex. For instance, in Albayrak and Allahverdi (2011) a greedy sub tour mutation operator is presented which is based on six parameters which control it. Here we see an idea frequently used in metaheuristics: do not perform just random changes but try to find local improvements as good as possible. This idea is also used in another suggestion for a crossover given in Ahmed (2010) which is denoted as sequential constructive crossover (SCX). Here, offspring chromosomes are constructed from parent chromosomes by trying to keep the original order but looking for best nodes to attach to a tour. Note that this kind of crossover does not use any random mechanisms. Other more complex operators could be based on a local search such as considered, for instance, in the concept of memetic algorithms. For instance, such embedded local search approaches could use a 2-opt approach or a Lin-Kernighan heuristic.

3.4.3 *Other Metaheuristics and Neural Networks for the Travelling Salesman Problem*

Some other metaheuristics like ant colony optimization (ACO) can be applied to a TSP in a rather straightforward manner: In ACO all possible connections in a tour (i.e. all edges in a respective graph) are labeled by values representing the pheromone levels which indicate the preferability of using a respective edge. During the run of the algorithm these pheromone levels change, being reinforced when the respective edge is used by an ant, otherwise decreased due to pheromone evaporation. An actual tour (corresponding to an ant in the ant colony) is then created step-by-step by using a probabilistic function which determines the preferability of using respective edges between locations during the tour. The probabilistic function favors strong pheromone levels on edges and short distances between locations but also allows occasionally for less favorable connections due to its stochastic nature.

Other metaheuristics require stronger adaptations or use modifications similar to those of evolutionary approaches. For instance, in the case of the usual real-number based particle swarm optimization (PSO), adaptations are required to consider the combinatorial nature of the TSP. For instance, in Shi et al. (2007), a representation of solutions as permutations is used which require further adaptations of the PSO operators, e.g. the consideration of differences of solutions as number of transpositions and the movement of solutions is specified by a slide operator. Moreover, specific crossover operators are used in this approach. In another PSO-based approach by Zhong et al. (2007), $n \times n$-matrices corresponding to all edges between locations are used for the solution representation, or to be more specific: to represent probabilities of using particular connections. These real-valued solution representations can be used for a rather straight-forward application of the standard PSO concepts.

The adaptation requirements for some other metaheuristics may be limited. For instance, in their study on using simulated annealing and similar metaheuristics for the TSP Pepper et al. (2002) only need to reconsider the step how new solutions are generated. Instead of using fully random neighbors of the current solutions, they apply a two-exchange procedure but limit possible exchanges to those of the pre-calculated nearest neighbors of each node.

Apart from evolutionary algorithms, various other metaheuristics have been applied frequently to the TSP and variants of it:

- ant colony optimization (Dorigo and Gambardella 1997a, b)
- GRASP (Marinakis et al. 2005)
- particle swarm optimization (Li et al. 2006)

- simulated annealing, threshold accepting and similar algorithms (Pepper et al. 2002)
- variable neighborhood search (Mladenović et al. 2012)
- memetic algorithms (Buriol et al. 2004; Ulder et al. 1991)

Also less well-known or very recent contributions to metaheuristics have been employed, e.g.:

- artificial bee colony algorithms (Karaboga and Gorkemli 2011)
- cuckoo search (Ouaarab et al. 2014)
- evolutionary ant rules (Tsai and Tsai 2002)
- the firefly algorithm (Jati and Suyanto 2011)
- jump search (Tsubakitani and Evans 1998),
- neuro-immune networks (Pasti and De Castro 2006)
- the noising method (Charon and Hudry 2000)

Also neural networks have been used repeatedly for solving TSP instances (see surveys in Potvin 1993 and Steinhaus 2015). In particular, Kohonen networks [also denoted as self-organizing maps (SOM) or self-organizing feature maps (SOFM)], have become popular because their data structures may represent geographical locations and their distances. In Kohonen networks all neurons are mutually connected and associated with Euclidean coordinates which are adapted during the learning process of the neural network. For the representation of a tour building problem with n locations, there should be a corresponding network with $m \geq n$ nodes. Typically, the number of neurons m in the network is clearly larger than n, e.g. $m = 2n$ (Créput and Koukam 2009). In Kohonen networks, the dynamics of the network finally leads to a solution which represents a tour by associating some of the neurons to destinations based on geometrical closeness. Besides SOM approaches, the similar concept of elastic nets is occasionally applied to respective tour building problems. Also, Hopfield neural networks (HNN), a type of fully connected feedforward networks, can be used for representing a transportation network and may help to derive a suitable tour (Jolai and Ghanbari 2010).

3.4.4 *On the Performance of Solution Approaches*

In general, it is difficult to judge which of the considered methods (and other approaches) are best for solving the TSP despite the relatively good basis of test problems. This is, for instance, because of difficulties in comparing run times under different conditions or with respect to specific method details and parameter settings. In many studies there are comparisons or benchmark results with respect to other methods but the computational studies are not sufficiently clear in experimental details. Moreover, usually a bias can be assumed due to an intensive tuning (of parameters, operators, etc.) of an author's newly suggested methodology whereas an existing method may be used without such tuning or with some arbitrarily chosen settings which may not be very suitable for the problem under consideration.

However, let us report a few comparative results: In a direct comparison of a neural network similar to Hopfield networks, an elastic net approach, a genetic algorithm, simulated annealing, and a hybrid approach based on simulated annealing and exhaustive local search, the genetic algorithm performed best in an older study by Peterson (1990), followed by the hybrid approach, the elastic net, and the other neural network while simulated annealing performed worst. In the more recent comparison by Stützle et al. (2000) the following methods have been compared: a genetic algorithm with a distance-preserving crossover (DPX), a repair-based genetic algorithm, ant colony optimization, iterated local search with fitness-distance-based diversification (ILS-FDD). All algorithms make use of a special 3-opt heuristic for local search. While all of the algorithms achieved very good results, ILS-FDD was best for most large-size test problems. In the comparative study by Weise et al. (2014), it was found that "local search can outperform pure global search methods on the TSP, but that [Evolutionary Computation] methods hybridized with local search can be even better." A population-based ant colony optimization (PACO) and a local search-based approach called Multi-Neighborhood Search (MNS) were found as best approaches in the respective tests on smaller size TSP instances.

3.5 The Vehicle Routing Problem

The Vehicle Routing Problem (VRP), also denoted as capacitated vehicle routing problem (CVRP), is one of the most popular transportation problem formulations due to its more realistic assumptions. Some good surveys on the VRP and variants are given in Anbuudayasankar et al. (2014), Golden et al. (2008), Liong et al. (2008), and Toth and Vigo (2014).

The VRP assumes that a number of locations are to be visited during a tour (just as in the case of the TSP) but capacity constraints of the vehicles need to be considered. A typical scenario where this is important is the delivery of goods to customers or the collection of goods at different locations (e.g. the collection of letters or parcels) to bring them to a common destination.

In its basic form the VRP can be formulated as follows for the case of delivering goods from a depot: Given a capacity q of the vehicle and n transport orders to the locations $l_1, l_2, \ldots, l_n$, each with goods requiring a capacity of a_i, $i = 1, \ldots, n$, one or several sequences of locations are to be found which serve all transport orders and respect the capacity of the vehicle.

As in the case of the TSP a route starts and ends in a given location which is usually called the depot (of the vehicles). The objective of the problem is usually to find several routes (or round trips) as short (or inexpensive) as possible.

Mathematically the problem can be formulated as an integer optimization problem (see, e.g., Kallehauge et al. 2005):

$$\min c(X) = \sum_{k \in V} \sum_{i \in N} \sum_{j \in N} c_{ij} x_{ijk} \tag{3.6}$$

such that

$$\sum_{k \in V} \sum_{j \in N} x_{ijk} = 1, \text{ for all } i \in \{1, \ldots, n\}, \tag{3.7}$$

$$\sum_{i \in N-\{0\}} a_i \sum_{j \in N} x_{ijk} \leq q, \text{ for all } k \in V, \tag{3.8}$$

$$\sum_{j \in N} x_{0jk} = 1, \text{ for all } k \in V, \tag{3.9}$$

$$\sum_{i \in N} x_{ihk} - \sum_{j \in N} x_{hjk} = 0, \text{ for all } h \in \{1, \ldots, n\}, k \in V, \tag{3.10}$$

$$\sum_{i \in N} x_{i,n+1,k} = 1, \text{ for all } k \in V, \tag{3.11}$$

$$x_{ijk} \in \{0, 1\} \text{ for all } i, j \in N, k \in V, \tag{3.12}$$

The decision variables are x_{ijk} defined in (3.12) as binary. These variables specify whether vehicle k visits location j after location i. In that case $x_{ijk} = 1$, otherwise it is 0. It is assumed that there is a set V of vehicles available and the set of locations N is expressed by respective index values, i.e. $N = \{0, \ldots, n\}$ with 0 indicating the depot.

In the problem formulation (3.6) specifies the objective of minimizing the total travel cost. Equation (3.7) ensures that that each customer is visited exactly once. The capacity limit of the vehicles is expressed by constraint (3.8). Equation (3.9) makes sure that each vehicle leaves the depot while (3.10) guarantees that each vehicle leaves a customer where it has arrived just before in the tour. Constraint (3.11) expresses the fact that the vehicles should return to the depot at the end of the tour.

Note that the adherence of the capacity is independent of the sequence of nodes inside one route. We only need to check that each single tour serves customers with added capacity requirements being smaller than the available capacity. Let us also mention that the problem without constraint (3.8) is a mathematical formulation of the travelling salesman problem.

Usually, the CVRP is interpreted as the problem of building several round trips for several vehicles (each for one of the vehicles) which are assumed to be identical with respect to capacity.

We could also interpret the problem as building several round trips for one vehicle (or a smaller number of vehicles than the required number of tours). This single vehicle would possibly need to return to its starting location (when not all goods fit into the vehicle due to the capacity limit) and do a second roundtrip (multitrip VRP). The problem formulation would not change, only the tours would need to be executed in a sequential, non-overlapping fashion instead of possibly concurrent tours of several vehicles. The different interpretation would only be relevant if further aspects are to be observed (e.g. time windows, see below).

3.5.1 The Vehicle Routing Problem with Time Windows

A very common variant of the vehicle routing problem requires considering additionally time windows for the deliveries of goods at the planned destinations of the transportation orders. That means, for each customer i it is required that the goods arrive within a pre-specified time interval $[s_i^{\min}, s_i^{\max}]$. If time windows are required only for some of the customers, we can still use the general formulation with assigning a time window $[-\infty,\infty]$ for customers without a time window.

The time windows can be expressed in the mathematical optimization problem (3.6)–(3.12) by two additional constraints:

$$x_{ijk}\left(s_{ik} + t_{ij} - s_{jk}\right) \leq 0 \text{ for all } i,j \in N, k \in V, \tag{3.13}$$

$$s_i^{\min} \leq s_{ik} \leq s_i^{\max} \text{ for all } i \in N, k \in V, \tag{3.14}$$

Constraint (3.13) defines arrival (and departure) times of vehicles at particular destinations and establishes the relationships between the vehicle departure times from a customer and its immediate successor. Constraint (3.14) makes sure that the time windows are observed. Note that the nonlinear constraint (3.13) can be linearized in order to make it better manageable for standard optimization techniques (Kallehauge et al. 2005).

3.5.2 The Vehicle Routing Problem with Multiple Vehicles

As mentioned above, for the VRP it is usually not distinguished whether the multiple routes that are calculated are executed by one or several identical vehicles (homogenous vehicles), because the feasibility and objective values (in the case of route length) would always be identical with both interpretations, although in the first case the tours would have to be operated in sequence by a single vehicle whereas the transports could be conducted in parallel by several vehicles in the second case. This situation might become different when further aspects are to be considered, e.g. time windows as discussed above, or when the vehicles are different and may have, in particular, different capacities (heterogeneous vehicles).

In general, a situation with explicit consideration of multiple vehicles which can carry out the transportation orders requires defining routes for each vehicle which cover the assigned orders. For each vehicle and the assigned orders it needs to be checked whether the capacities are met.

Let us note that such types of problems considering a fleet of vehicles with heterogeneous capacities (and possibly further equipment) appear to be more realistic than the standard VRP assumptions. This variant of the VRP can, of course, be formulated with time windows as well.

Vehicle routing problems with multiple vehicles may also consider different objectives. Instead of minimizing the total length of tours one may try to find the smallest number of vehicles so that their tours include all deliveries and respect the capacity limits.

3.5.3 The Vehicle Routing Problem with Multiple Depots

Another variant of the vehicle routing problem is to consider different depots for the vehicles. Usually, it is assumed that a transport order can be delivered from each depot. So it is necessary to assign each transport order to a depot from which it is going to be fulfilled and then to build respective tours. In the case of heterogeneous vehicles, it is necessary instead to assign transport orders to vehicles which are usually assumed to be assigned to a particular depot (as part of the problem formulation).

Instead of discussing a mathematical problem formulation we consider an example of a multi-depot vehicle routing problem as shown in Fig. 3.1. In the figure, there are three depots denoted by stars and 15 customers denoted by dots. In the solution shown, eight of the customers are served by the same depot during two tours while for each other depot only one tour is defined.

We could therefore state that the multi-depot vehicle routing problem (similar to the vehicle routing problem with heterogeneous vehicles) requires two tasks to be solved: assigning transport orders to specific depots (or individual vehicles associated with a specific depot) and building one or several tours for each depot. In fact, some solution approaches make use of this problem structure and solve the problem in a hierarchical way, first assigning specific resources (such as depots or vehicles) to the transport orders and then building a tour for each resource.

Even if no such two-phase approach is applied, both types of information must be encoded in a respective method. For instance, in the GA-based approach Li et al. (2016) use a single chromosome for representing a solution which consists of a part for each depot (with one gene indicating the depot) followed by the tour sequence for the depot. Let us note that sometimes the assignments part (also denoted as clustering) is not included in the solution approach. For instance, either a simple heuristic is applied a priori, e.g. assigning customers to the closest depot or, as in Liu et al. (2014), the assignment to a depot is done after the tours have been determined by a metaheuristic.

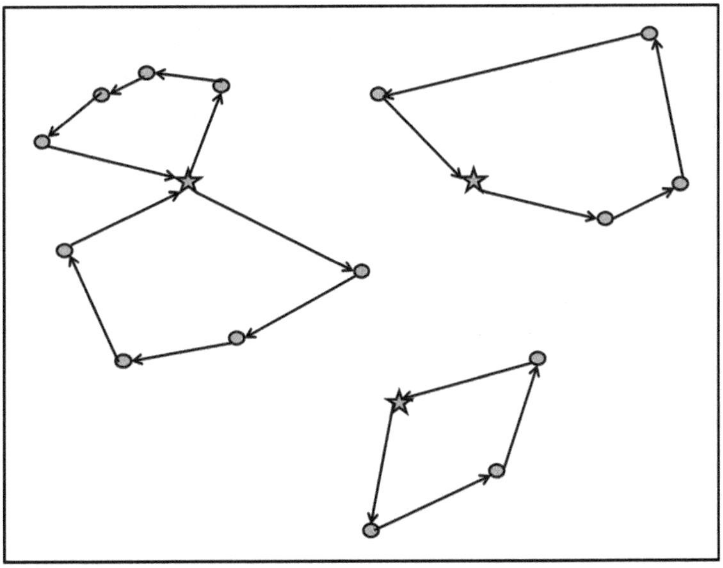

Fig. 3.1 Example of a multi-depot vehicle routing problem solution

3.5.4 More Differentiated Problem Variants

More complex variants of the VRP may consider further restrictions concerning feasible transports. For instance, capacities may not just be defined by scalar values for capacity requirements to be met by a number expressing the available capacity of a vehicle. Instead, the problem may assume defined 2-dimensionl or 3-dimensional characteristics of the loads to be placed into a respectively defined loading space of the vehicle. The problem is then a combination of a tour building problem and a packing problem (see, e.g. Leung et al. 2011).

Other problem formulations may consider the capacities as alternative loading forms of the vehicles. For instance, in Hanne et al. (2009) the transport of patients in clinic vehicles is considered which allow the transport of e.g. one recumbent patient and one sitting person or, alternatively, three sitting persons (e.g. patients with wheelchairs). Moreover, the further constraints concerning cargo consolidation may apply, e.g. restrictions on the mixing of different animal categories in livestock transportations, see, e.g., Oppen and Løkketangen (2006). Other examples of excluded cargo consolidation can be found in the transport of infectious patients or cargo requiring different handling or transport conditions, e.g. cooling.

Further practically relevant aspects may result from specific equipment required for the transport, e.g. ventilation or cooling devices which necessitate either specific vehicles or the explicit modelling of additional devices to be assigned to transports. This may imply that from a fleet of vehicles not every vehicle is suitable despite offering sufficient capacity.

Another class of VRP variants which have received more attention in the recent past are multi-echelon vehicle routing problems and in particular the two-echelon vehicle routing problem (2E-VRP) which assumes that cargo is delivered from depots to intermediary satellites, from where it is then shipped to customers. Such problems may arise, for instance, when different vehicles are used for cost or capacity reasons or because of driving restrictions for large vehicles (e.g. for small roads in an old town city center). Some introduction into that kind of problems can be found in Perboli et al. (2011).

Finally, let us mention that, of course, various aspects of vehicle routing problems can be considered as uncertain, vague or varying over time. In their survey on dynamic and stochastic vehicle routing problems, Ritzinger et al. (2016) distinguish three main sources of uncertainty: stochastic travel times, stochastic demand, and stochastic customers. Frequently, situations with uncertainty or vagueness are modelled by fuzzy approaches. For an older survey on using fuzzy techniques for various kinds of problems in transport planning and control we refer to Teodorović (1999). More recent studies on fuzzy VRPs are presented, for instance, in Erbao and Mingyong (2009) with respect to fuzzy demand, in Xu et al. (2011) with respect to fuzzy time windows, and in Zheng and Liu (2006) with respect to fuzzy travel times.

3.6 Solution Approaches for Vehicle Routing Problems

As generalizations of the travelling salesman problem the capacitated vehicle routing problem and variants of it are NP-hard as well (Lenstra and Kan 1981). It is thus assumed that polynomial-time algorithms for solving such problems exactly do not exist and a lot of effort has been devoted to finding heuristics and metaheuristics which should produce sufficiently good solutions. The quality of algorithms is usually measured in experiments using test problems. Respective test problems for the VRP and some other extensions of the TSP are also available in the mentioned TSPLIB library. Further test problems are discussed and used in several publications. In particular, the set of test problems for VRPTW, provided by Solomon (2014) is used more often. Another set of test problems is included in the OR Library by Beasley (2014). However, the general coverage of these problems by standard benchmark problems is not done as thoroughly as for the TSP.

As for other optimization problems we can roughly distinguish three types of methods: simple heuristics, metaheuristics, and mathematical approaches. A good overview of simple heuristics is given by Laporte and Semet (2001) who distinguish constructive heuristics, two-phase heuristics consisting of a clustering phase (assignment of transport order to routes or vehicles) and a route building phase, and improvement heuristics. The improvement heuristics can be quite simple as the 2-opt (and generalizations) discussed above but may also consider the exchange of locations between different routes. With respect to mathematical approaches, typical techniques are similar to those for other hard combinatorial problems, e.g. column generation, branch-and-bound, or branch-and-price approaches. In the following, we consider the application of metaheuristics in more detail.

As for the TSP, solutions for the VRP and variants can be represented as a sequence of locations (or their respective indices). Simple variation operators such as mutation and crossover in evolutionary algorithms must take care that a possibly random change of a sequence does not lead to leaving out a location or to the double consideration of a location. Therefore, a typical change would be based on the exchange of at least two locations in a sequence (see Sect. 3.4.2). Other restrictions such as capacity limits or time windows require additional consideration in such changes to maintain the feasibility of solutions.

In general, we can remark that respective adaptations of making CI approaches applicable to VRP problems including solution representations are quite similar to those discussed above for the TSP. In particular, a similarly large number of different metaheuristics as for the TSP has been applied to the VRP and variants.

In their categorized bibliography on using metaheuristics for different types of VRPs (standard VRP, VRP with time windows, VRP with backhauls, pickup and delivery problems, VRP with multiple use of vehicles, with mixed fleet, with multiple depots and periodic problems, and dynamic VRPs), Gendreau et al. (2008) consider the following numbers of publications for different types of metaheuristics (with possible double counting if an approach is used for different problem types):

- tabu search: 65
- genetic algorithms: 43
- ant colony optimization: 10
- simulated annealing: 10
- variable neighborhood search: 4
- greedy randomized adaptive search procedure: 3
- others: 51

Apart from the impressive overall number of metaheuristics-based approaches it is noticeable, that the large number of other metaheuristics reflects the ongoing search for new solution approaches in that area, most of them used so far in only one or two publications. Some examples of these metaheuristics are particle swarm optimization, adaptive large neighborhood search iterated local search, path relinking, but we also find a few newer and less known metaheuristic approaches such as a heuristic bubble algorithm or an imperialist competitive algorithm. Often, the applied techniques are memetic algorithms or other hybrid approaches.

Results comparable to those of Gendreau et al. (2008) are presented in a survey by Jozefowiez et al. (2008) on multiobjective vehicle routing problems: Evolutionary algorithms appear as most often used approaches besides some types of simple heuristics. Other frequently used approaches are various types of local search heuristics and ant colony optimization while other approaches are used more sporadically. The multiobjective aspect is usually either treated by weighting the objectives, by a goal programming formulation or by an approximation of the Pareto set (as often done in multiobjective evolutionary algorithms).

Also neural networks have been used occasionally for dealing with vehicle routing problems. Compared to their application to the TSP relating studies are less frequent which is probably due to the fact that further restrictions are difficult to embed in a neural network-based model. Among the respective studies we find Modares et al. (1999) who apply a Self-Organizing Map (Kohonen network) to such problems which is more successful than some of the other analysed neural network variants. Similar approaches are used in Steinhaus et al. (2015) and Steinhaus (2015). The capacity constraint is here considered by assigning neurons to different rings corresponding to separate tours. Moreover, the approach uses an improved bias term and makes use of a novel restricted mechanism for increasing the feasibility of solutions. A dynamic VRP has also been solved by a Self-Organizing Map approach which has been hybridized with an evolutionary algorithm (Créput et al. 2012). In a series of experiments the resulting algorithm is shown to outperform a genetic algorithm-based method and an ant colony approach.

With respect to the performance of different types of algorithms, general statements are more difficult than for the TSP due to a smaller standardization of the problem and a less well-established canon of test problems. Let us nevertheless present some comparative results from the literature: Evolutionary algorithms seem to be the dominating complex metaheuristics, but many studies have shown that good design and problem-specific adaptations are key factors for superior performance.

Dorigo and Gambardella (1997a, b) showed a competitive or superior performance of ACO compared to other metaheuristics. For a number of mid-sized test problems optimal results could be derived. The development of neural networks on the other hand remained stuck towards competitive approaches despite some promising first results. Only during recent years these concepts have been revisited with some success. Very good results have also been obtained for hybrid metaheuristics, especially various metaheuristics which were combined with local search (see, e.g., Nguyen et al. 2007), e.g. memetic algorithms which combine an evolutionary algorithm with local search for improving newly created intermediate solutions. Typically, for the local search simple heuristics such as 2-opt, 3-opt, or the more complex Lin-Kernighan improvement operations have been used. On the one hand, such improvement operations are very effective even when used alone. On the other hand, metaheuristic schemes may incorporate better perturbation mechanisms than the usually random kicks used in established Lin-Kernighan methods for leaving a local optimum. That means, the random mechanisms of an evolutionary algorithm and a diversified population may allow more effectively to escape from local optima or local basins.

3.7 The Pickup and Delivery Problem

The pickup and delivery problem is another common type of transportation problem. Unlike for the VRP not only delivery or pickup activities are to be planned within the route but both activities. That means, the vehicle usually starts empty at some depot and needs to pick up goods at different locations along its way (as specified by given transport orders) and needs to bring them to the respective destinations. This is the typical business case of a transport service provider which transports goods for various customers while the standard VRP rather considers transport activities conducted by a company on own account, e.g. distribution of goods from a warehouse which also serves as depot for own vehicles. A special case of a pickup and delivery problem is when the transport service provider picks up goods only at one location which are to be transported to several customers. In that case, the route starts at the depot of the service providers, then goes to the client location, then serves the customers in a useful sequence, and finally returns to the depot.

As in the case of VRP, pickup and delivery problems usually require to observe the capacity constraints of the vehicle. That could imply that some locations are to be visited twice, for instance, a location where goods for different customers are to be picked up. This is the case when not all goods which have to be picked up at that place fit into the vehicle simultaneously. Just like in the case of the capacitated VRP, it is often necessary to observe time windows: now possibly for both pickup and delivery locations.

Let us consider an example with the matrix of travel times being given in Table 3.2. A transport service provider, whose vehicle depot is in D, has to pick up shipments at several locations and bring them to different locations. For loading and unloading at any location 10 min are generally assumed.

The following transports are to be executed:

1. a shipment from A to B,
2. a shipment from C to B,
3. a shipment from C to A.

If the capacity of vehicles is sufficient to transport these shipments simultaneously we can use the following consideration for simplifying the problem: If possible, each location should be visited only once. Here, this means that the two shipments from C should be picked up during the same stop, for A the pickup of the first shipment and the unloading of the third shipment should happen during the

Table 3.2 Example distance matrix

	A	B	C	D
A		12	35	25
B	12		25	10
C	35	25		20
D	25	10	20	

same stop and both shipments for B should be delivered together. The following “natural” tour results:

$D_{departure}0'$ $C_{arrival}20'$ $C_{departure}30'$ $A_{arrival}65'$ $A_{departure}75'$ $B_{arrival}87'$ $B_{departure}97'$ $D_{arrival}107'$

If, for instance, shipment 2 requires the complete capacity of the vehicle this solution is no longer feasible. We can then assume that two partial tours $C \rightarrow B$ and $C \rightarrow A \rightarrow B$ must be combined in the complete tour starting and ending in D. Thus, two equally good solutions result:

$D_{departure}0'$ $C_{arrival}20'$ $C_{departure}30'$ $B_{arrival}55'$ $B_{departure}65'$ $C_{arrival}90'$$C_{departure}100'$ $A_{arrival}135'$$A_{departure}145'$ $B_{arrival}157'$ $B_{departure}167'$ $D_{arrival}177'$

$D_{departure}0'$ $C_{arrival}20'$ $C_{departure}30'$ $A_{arrival}65'$ $A_{departure}75'$ $B_{arrival}87'$ $B_{departure}97'$ $C_{arrival}122'$$C_{departure}132'$ $B_{arrival}157'$ $B_{departure}167'$ $D_{arrival}177'$

If we assume that the tour starts at 8:00 and a time window from 9:15 to 9:45 has to be observed for the pickup of goods at A then the first tour would no longer be feasible because the respective pickup takes place between 10:15 and 10:25. The second solution can be realized if a waiting time of 10’ at A is considered (waiting from 9:05 arrival time until 9:15, the beginning of the time window). The whole tour would then be delayed by this 10’ and end at 11:07.

If otherwise a time window from 8:30 to 9:00 for the pickup of goods at C should be observed, there would be no feasible solution as C needs to be visited twice and none of the respective deliveries can be accomplished within 30’ including return to C. The conflict could be resolved by changing the problem formulation, e.g. by using two vehicles, by allowing for a temporary storage of one shipment close to C, or by agreeing on a wider time window.

Let us note that also other variants are considered in the literature on pickup and delivery problems similar to VRPs, e.g. multiple vehicles, specific constraints on loading or required devices, or problems with uncertainty or vague information. A special variant of pickup and delivery problems has attracted more attention in the literature. This problem denoted as dial-a-ride problem requires an additional constraint that the transport duration of each load is limited (maximum duration is specified). The constraint shall prevent solutions where between the pickup and the delivery of a good too many other pickups and deliveries are planned within a tour. Such limitations on transport durations may be relevant when humans are transported or may even be based on legal restrictions, e.g. in the case of livestock transportation.

3.8 Network Flow Problems

Another type of transportation problem is based on the assumption of optimizing the flows of goods in a network. The network is represented by a directed (or undirected) graph. Transportation demand is specified by network nodes with demand (sinks) and network nodes with the supply of goods (sources).

Network flow problems exist in two basic variants with respect to the restrictions and objective function. One variant assumes that the capacity of edges in the graph is limited, i.e. $f(u,v) \leq c(u,v)$ where f is the flow on an edge $(u.v)$ and $c(u,v)$ is the respective capacity limit. The objective here is to determine the maximum load which can be shipped through the network from a starting location a to a destination b. Assuming that all amounts to be transported through the various edges are real variables, the problem can be formulated as a linear optimization problem and can be solved with established methods. A frequently considered extension of the basic problem is the multicommodity-flow problem (MCFP) (see, e.g., McBride 1998 for a survey). Here different types of materials which have to be transported through the network are considered. For all these problems the specific transport activities (e.g. the vehicles to be used) are not taken into account. Moreover, the solution may require that a large number of edges is used for the transportation task with the consequence that the freight needs to be split up and is transported with different vehicles along different routes. This might lead to rather long distances and high transport costs. From the computation complexity point of view, let us mention, that real-valued problems can be solved in polynomial time while integer variants are shown to be NP-hard (Even et al. 1975; Hall et al. 2007). Unlike in most other considered transport planning problems, metaheuristics play yet a minor role in solving network flow problems which mostly attract solution approaches from the field of mathematical programming.

Another type of network flow problem might arise when the transportation problem is considered from the viewpoint of a company which wants to use transport service providers for conducting the transports. Using modern IT technologies, it has become rather simple to obtain transport costs electronically from respective service providers. The transport service user is not interested in the specific execution of the transports (such as used vehicles) but can inquire for respective cost information and further information such as, e.g., respective capacities or transport times. Possibly, further practically relevant aspects are to be observed, e.g. time windows, cargo consolidation requirements, crossdocking, or scheduling aspects. The objective is usually to minimize some kind of cost function. A related approach is presented by Bryan and McIntire (2015) who use a cloud-based architecture which employs different types of metaheuristics in an agent-based framework.

References

Ahmed, Z. H. (2010). Genetic algorithm for the traveling salesman problem using sequential constructive crossover operator. *International Journal of Biometrics and Bioinformatics (IJBB), 3*(6), 96–105.

Albayrak, M., & Allahverdi, N. (2011). Development a new mutation operator to solve the traveling salesman problem by aid of genetic algorithms. *Expert Systems with Applications, 38*(3), 1313–1320.

Anbuudayasankar, S. P., Ganesh, K., & Mohapatra, S. (2014). *Models for practical routing problems in logistics*. Cham: Springer.

Applegate, D., Cook, W., & Rohe, A. (2003). Chained Lin-Kernighan for large traveling salesman problems. *INFORMS Journal on Computing, 15*(1), 82–92.

Beasley, J. E. (2014). *OR-library*. Retrieved March 12, 2016, from http://people.brunel.ac.uk/~mastjjb/jeb/info.html

Braun, H. (1991). On solving travelling salesman problems by genetic algorithms. In H.-P. Schwefel & R. Männer (Eds.), *Parallel problem solving from nature* (pp. 129–133). Berlin and Heidelberg: Springer.

Bryan, J., & McIntire, J. (2015). *Leveraging distributed metaheuristics to balance flexibility and complexity*. Invited Talk at IEEE TENCON 2015: Technical Track 11, IEEE, New York.

Buriol, L., França, P. M., & Moscato, P. (2004). A new memetic algorithm for the asymmetric traveling salesman problem. *Journal of Heuristics, 10*(5), 483–506.

Burkard, R. E., & Cela, E. (1999). Linear assignment problems and extensions. In D.-Z. Du & P. M. Pardalos (Eds.), *Handbook of combinatorial optimization* (pp. 75–149). New York: Springer.

Charon, I., & Hudry, O. (2000). Application of the noising method to the travelling salesman problem. *European Journal of Operational Research, 125*(2), 266–277.

Chatterjee, S., Carrera, C., & Lynch, L. A. (1996). Genetic algorithms and traveling salesman problems. *European Journal of Operational Research, 93*(3), 490–510.

Créput, J. C., Hajjam, A., Koukam, A., & Kuhn, O. (2012). Self-organizing maps in population based metaheuristic to the dynamic vehicle routing problem. *Journal of Combinatorial Optimization, 24*(4), 437–458.

Créput, J. C., & Koukam, A. (2009). A memetic neural network for the Euclidean traveling salesman problem. *Neurocomputing, 72*(4), 1250–1264.

Crisan, G. C., & Nechita, E. (2008). Solving fuzzy TSP with ant algorithms. *International Journal of Computers, Communications and Control, 3*, 228–231.

Dorigo, M., & Gambardella, L. M. (1997a). Ant colonies for the travelling salesman problem. *BioSystems, 43*(2), 73–81.

Dorigo, M., & Gambardella, L. M. (1997b). Ant colony system: a cooperative learning approach to the traveling salesman problem. *IEEE Transactions on Evolutionary Computation, 1*(1), 53–66.

Erbao, C., & Mingyong, L. (2009). A hybrid differential evolution algorithm to vehicle routing problem with fuzzy demands. *Journal of Computational and Applied Mathematics, 231*(1), 302–310.

Even, S., Itai, A., & Shamir, A. (1975). *On the complexity of time table and multi-commodity flow problems*. 16th Annual Symposium on Foundations of Computer Science, 1975, IEEE, New York, pp. 184–193.

Feillet, D., Dejax, P., & Gendreau, M. (2005). Traveling salesman problems with profits. *Transportation Science, 39*(2), 188–205.

Fu, L., Sun, D., & Rilett, L. R. (2006). Heuristic shortest path algorithms for transportation applications: state of the art. *Computers & Operations Research, 33*(11), 3324–3343.

Gendreau, M., Laporte, G., & Vigo, D. (1999). Heuristics for the traveling salesman problem with pickup and delivery. *Computers & Operations Research, 26*(7), 699–714.

Gendreau, M., Potvin, J. Y., Bräumlaysy, O., Hasle, G., & Løkketangen, A. (2008). Metaheuristics for the vehicle routing problem and its extensions: A categorized bibliography. In B. L. Golden, S. Raghavan, & E. A. Wasil (Eds.), *The vehicle routing problem: Latest advances and new challenges* (pp. 143–16). New York: Springer.

Golden, B. L., Raghavan, S., & Wasil, E. A. (Eds.). (2008). *The vehicle routing problem: latest advances and new challenges* (Operations Research/Computer Science Interfaces Series, Vol. 43). New York: Springer.

Grefenstette, J., et al. (1985). *Genetic algorithms for the traveling salesman problem.* Proceedings of the first international conference on Genetic Algorithms and Their Applications, Lawrence Erlbaum, New Jersey, pp. 160–168.

Hall, A., Hippler, S., & Skutella, M. (2007). Multicommodity flows over time: Efficient algorithms and complexity. *Theoretical Computer Science, 379*(3), 387–404.

Hanne, T., Melo, T., & Nickel, S. (2009). Bringing robustness to patient flow management through optimized patient transports in hospitals. *Interfaces, 39*(3), 241–255.

Jati, G. K., & Suyanto, S. (2011). Evolutionary discrete firefly algorithm for travelling salesman problem. In *Adaptive and intelligent systems.* Lecture Notes in Computer Science (Vol. 6943, pp. 393–403). Berlin and Heidelberg: Springer.

Jolai, F., & Ghanbari, A. (2010). Integrating data transformation techniques with Hopfield neural networks for solving travelling salesman problem. *Expert Systems with Applications, 37*(7), 5331–5335.

Jozefowiez, N., Semet, F., & Talbi, E. G. (2008). Multi-objective vehicle routing problems. *European Journal of Operational Research, 189*(2), 293–309.

Kallehauge, B., Larsen, J., Madsen, O. B., & Solomon, M. M. (2005). Vehicle routing problem with time windows. In G. Desaulniers, J. Desrosiers, & M. M. Solomon (Eds.), *Column generation* (pp. 67–98). New York: Springer.

Karaboga, D., Gorkemli, B. (2011). *A combinatorial artificial bee colony algorithm for traveling salesman problem.* 2011 international symposium on Innovations in Intelligent Systems and Applications (INISTA), IEEE, Piscataway, NJ, pp. 50–53.

Laporte, G., & Semet, F. (2001). Classical heuristics for the capacitated VRP. In P. Toth & D. Vigo (Eds.), *The vehicle routing problem* (pp. 109–128). Philadelphia: Society for Industrial and Applied Mathematics.

Larrañaga, P., Kuijpers, C. M. H., Murga, R. H., Inza, I., & Dizdarevic, S. (1999). Genetic algorithms for the travelling salesman problem: A review of representations and operators. *Artificial Intelligence Review, 13*(2), 129–170.

Lenstra, J. K., & Kan, A. H. G. (1981). Complexity of vehicle routing and scheduling problems. *Networks, 11*(2), 221–227.

Leung, S. C., Zhou, X., Zhang, D., & Zheng, J. (2011). Extended guided tabu search and a new packing algorithm for the two-dimensional loading vehicle routing problem. *Computers & Operations Research, 38*(1), 205–215.

Li, J., Li, Y., & Pardalos, P. M. (2016). Multi-depot vehicle routing problem with time windows under shared depot resources. *Journal of Combinatorial Optimization, 31*(2), 515–532.

Li, X., Tian, P., Hua, J., & Zhong, N. (2006). A hybrid discrete particle swarm optimization for the traveling salesman problem. In T.-D. Wang, X. Li, & X. Wang (Eds.), *Simulated evolution and learning* (pp. 181–188). Berlin and Heidelberg: Springer.

Liong, C. Y., Wan, R. I., Khairuddin, O., & Zirour, M. (2008). Vehicle routing problem: Models and solutions. *Journal of Quality Management and Analysis, 4*, 205–218. Retrieved March 12, 2016, from http://pkukmweb.ukm.my/~ppsmfst/jqma/Vol4_Is1/PDF/JQMA4(1)-19-liong-drk.pdf

Liu, T., Jiang, Z., & Geng, N. (2014). A genetic local search algorithm for the multi-depot heterogeneous fleet capacitated arc routing problem. *Flexible Services and Manufacturing Journal, 26*(4), 540–564.

Marinakis, Y., Migdalas, A., & Pardalos, P. M. (2005). Expanding neighborhood GRASP for the traveling salesman problem. *Computational Optimization and Applications, 32*(3), 231–257.

McBride, R. D. (1998). Advances in solving the multicommodity-flow problem. *Interfaces, 28*(2), 32–41.

Mladenović, N., Urošević, D., & Ilić, A. (2012). A general variable neighborhood search for the one-commodity pickup-and-delivery travelling salesman problem. *European Journal of Operational Research, 220*(1), 270–285.

Modares, A., Somhom, S., & Enkawa, T. (1999). A self-organizing neural network approach for multiple traveling salesman and vehicle routing problems. *International Transactions in Operational Research, 6*(6), 591–606.

Nguyen, H. D., Yoshihara, I., Yamamori, K., & Yasunaga, M. (2007). Implementation of an effective hybrid GA for large-scale traveling salesman problems. *IEEE Transactions on Systems, Man, and Cybernetics, Part B (Cybernetics), 37*(1), 92–99.

Oppen, J., & Løkketangen, A. (2006). *The livestock collection problem.* Proceedings of the 21st European conference on Operational Research.

Ouaarab, A., Ahiod, B., & Yang, X. S. (2014). Discrete cuckoo search algorithm for the travelling salesman problem. *Neural Computing and Applications, 24*(7–8), 1659–1669.

Pasti, R., & De Castro, L. N. (2006). *A neuro-immune network for solving the traveling salesman problem.* International joint conference on Neural Networks, 2006. IJCNN'06, IEEE, Piscataway, NJ, pp. 3760–3766.

Pepper, J. W., Golden, B. L., & Wasil, E. A. (2002). Solving the traveling salesman problem with annealing-based heuristics: a computational study. *IEEE Transactions on Systems, Man and Cybernetics, Part A: Systems and Humans, 32*(1), 72–77.

Perboli, G., Tadei, R., & Vigo, D. (2011). The two-echelon capacitated vehicle routing problem: Models and math-based heuristics. *Transportation Science, 45*(3), 364–380.

Peterson, C. (1990). Parallel distributed approaches to combinatorial optimization: Benchmark studies on traveling salesman problem. *Neural Computation, 2*(3), 261–269.

Potvin, J. Y. (1993). State-of-the-art survey—the traveling salesman problem: A neural network perspective. *ORSA Journal on Computing, 5*(4), 328–348.

Reinelt, G. (1991). TSPLIB—a traveling salesman problem library. *ORSA Journal on Computing, 3*(4), 376–384.

Ritzinger, U., Puchinger, J., & Hartl, R. F. (2016). A survey on dynamic and stochastic vehicle routing problems. *International Journal of Production Research, 54*(1), 215–231.

Shi, X. H., Liang, Y. C., Lee, H. P., Lu, C., & Wang, Q. X. (2007). Particle swarm optimization-based algorithms for TSP and generalized TSP. *Information Processing Letters, 103*(5), 169–176.

Shim, V. A., Tan, K. C., Chia, J. Y., & Chong, J. K. (2011). Evolutionary algorithms for solving multi-objective travelling salesman problem. *Flexible Services and Manufacturing Journal, 23* (2), 207–241.

Snyder, L. V., & Daskin, M. S. (2006). A random-key genetic algorithm for the generalized traveling salesman problem. *European Journal of Operational Research, 174*(1), 38–53.

Solomon, M. M. (2014). *VRPTW benchmark problems.* Retrieved March 12, 2016, from http://w.cba.neu.edu/~msolomon/problems.htm

Steinhaus, M. (2015). *The application of the self organizing map to the vehicle routing problem.* Open Access Dissertations, University of Rhode Island, Kingston.

Steinhaus, M., Shirazi, A. N., & Sodhi, M. (2015). *Modified self organizing neural network algorithm for solving the vehicle routing problem.* 2015 I.E. 18th International Conference on Computational Science and Engineering (CSE), IEEE, Piscataway, NJ, pp. 246–252.

Stützle, T., Grün, A., Linke, S., & Rüttger, M. (2000). A comparison of nature inspired heuristics on the traveling salesman problem. In M. Schoenauer, K. Deb, G. Rudolph, X. Yao, E. Lutton, J. J. Merelo, & H.-P. Schwefel (Eds.), *Parallel problem solving from nature PPSN VI* (pp. 661–670). Berlin and Heidelberg: Springer.

Takaoka, T. (2013). A simplified algorithm for the all pairs shortest path problem with $O(n^2 \log n)$ expected time. *Journal of Combinatorial Optimization, 25*(2), 326–337.

Teodorović, D. (1999). Fuzzy logic systems for transportation engineering: the state of the art. *Transportation Research Part A: Policy and Practice, 33*(5), 337–364.

Toth, P., & Vigo, D. (Eds.). (2014). *Vehicle routing: Problems, methods, and applications* (Vol. 18). Philadelphia: Society for Industrial and Applied Mathematics.

Tsai, C. F., Tsai, C. W. (2002). *A new approach for solving large traveling salesman problem using evolutionary ant rules*. Proceedings of the 2002 international joint conference on Neural Networks, 2002 (IJCNN'02), Vol. 2, IEEE, Piscataway, NJ.

Tsai, H. K., Yang, J. M., Tsai, Y. F., & Kao, C. Y. (2004). An evolutionary algorithm for large traveling salesman problems. *IEEE Transactions on Systems, Man, and Cybernetics, Part B: Cybernetics, 34*(4), 1718–1729.

Tsubakitani, S., & Evans, J. R. (1998). Optimizing tabu list size for the traveling salesman problem. *Computers & Operations Research, 25*(2), 91–97.

Ulder, N. L., Aarts, E. H., Bandelt, H. J., van Laarhoven, P. J., & Pesch, E. (1991). Genetic local search algorithms for the traveling salesman problem. In H.-P. Schwefel & R. Männer (Eds.), *Parallel problem solving from nature* (pp. 109–116). Berlin and Heidelberg: Springer.

Weise, T., Chiong, R., Lassig, J., Tang, K., Tsutsui, S., Chen, W., et al. (2014). Benchmarking optimization algorithms: An open source framework for the traveling salesman problem. *IEEE Computational Intelligence Magazine, 9*(3), 40–52.

Xu, J., Yan, F., & Li, S. (2011). Vehicle routing optimization with soft time windows in a fuzzy random environment. *Transportation Research Part E: Logistics and Transportation Review, 47*(6), 1075–1091.

Yadlapalli, S., Malik, W. A., Darbha, S., & Pachter, M. (2009). A Lagrangian-based algorithm for a multiple depot, multiple traveling salesmen problem. *Nonlinear Analysis: Real World Applications, 10*(4), 1990–1999.

Zheng, Y., & Liu, B. (2006). Fuzzy vehicle routing model with credibility measure and its hybrid intelligent algorithm. *Applied Mathematics and Computation, 176*(2), 673–683.

Zhong, W. L., Zhang, J., & Chen, W. N. (2007). *A novel discrete particle swarm optimization to solve traveling salesman problem*. IEEE Congress on Evolutionary Computation, 2007. CEC 2007, IEEE, Piscataway, NJ, pp. 3283–3287.

Chapter 4
Inventory Planning and Lot-Sizing

Abstract In this chapter we discuss problems in inventory planning and related planning tasks such as determining lot sizes for procurement or production. After a motivation of the importance of inventory planning we first discus basic models in inventory management such as the well-known equation for calculating economic order quantities. After that we discuss more complex lot-sizingsizing models and related solution approaches from computational intelligence. After that some more detailed problems in the planning of warehouse operations and questions concerning the determination of specific storage locations are treated. The last section of this chapter discusses some inventory routing models which combine the planning of inventory with the planning of transport tasks.

4.1 The Need for Inventory Planning

The management of inventories belongs to the core logistics activities. Storing materials close to the place where they are assumed to be needed at some future time can be advantageous, especially when their provision (e.g. by manufacturing or by transport) at the required time is too costly, too risky, or just not possible.

To be more specific, we can distinguish three main groups of reasons why goods are stored: aspects relating to procurement, in-house aspects, and aspects relating to the customers of a company. The procurement aspects relate to a company's interactions with suppliers. In particular, economic considerations usually lead to the procurement of larger quantities than are immediately needed. Buying larger quantities may reduce the price and may lead to fewer procurement activities and transportation processes which also should result in reduced procurement costs. Managing a stock of the procured items is a consequence of this consideration of such procurement-related costs.

Another supplier-related reason is the decrease of uncertainties in procurement: The ordered materials could arrive too late, or the quality could be insufficient, or the materials could arrive in too small quantities or not at all. For a company a way of dealing with such supply-related problems could be to have a sufficient amount

T. Hanne, R. Dornberger, *Computational Intelligence in Logistics and Supply Chain Management*, International Series in Operations Research & Management Science 244,
DOI 10.1007/978-3-319-40722-7_4

of the required material in stock. But even without such cost or risk considerations, a stock of the required items may be necessary when the lead time for procurement is undesirably high.

Another group of reasons for an inventory are the in-house operations of a company. It is very often the case that the production cannot be run in a single step or in a continuous process. As a result, there are intermediate products which require at least a short time of storage (buffering). Another reason may be the production technology which requires specific lot sizes such as minimum quantities of goods to be produced at one time. Minimum lot sizes may also result from economic considerations such as a decrease of production costs when there is a contiguous production of a larger quantity (economies of scale). A reason for this could be the avoidance of set-up times which occur when changing from one product to another. As a result, a company may have a larger quantity of a product available than is currently needed. The excess production is incorporated in the inventory.

Finally, there are several reasons which concern the customers of a company. First of all, in many industries, especially those producing consumer goods, the exact future demand is uncertain. If a company wants to have sufficient quantities of the product available to be always able to satisfy the uncertain customer demand it has to have an inventory of these products. Such an inventory is then also called a safety stock.

A related reason for an inventory could be the operation of warehouses close to the customers to enable a quick satisfaction of demand. Such warehouses may then also decrease related transportation costs because larger amounts can be transported to the areas of expected customers at smaller costs. Also a strongly seasonal demand of products may lead to an inventory. If, for instance, demand only occurs during short periods of a year, it might be preferable (because of cost reasons) to produce the respective goods not just shortly before those periods but to balance the production, e.g. produce the goods evenly over the year. In this way, stocks are being built during times of no demand and reduced during times of demand.

Other purposes of warehousing may include the specialization of products in the warehouse, e.g. attachment of country-specific cables, power supplies or manuals in the European central warehouse of a US IT manufacturer (van Hoek 2001). Also some goods may improve during their time of storage, e.g. in the case of cheese or wine (ripening processes). Another good reason for warehousing could be the expectation of higher prices in the future. Such speculative reasons may in particular apply for the warehousing of precious metals or artworks.

The expectation of higher prices or an improvement of the quality during warehousing is, however, an exception. More often, the quality deteriorates over time, e.g. in the case of most food products. In other cases, the quality may stay unchanged but due to technological progress or changed customer preferences people are not willing to buy the product anymore, or only at a lower price. For instance, computers or mobile phones which are a few years old are usually no longer sellable to customers.

The obsolescence of stored goods often leads to a significant amount of warehousing costs. Other warehousing costs result from the necessity of a physical warehouse which requires construction costs or respective write-offs or rents. Then there are the operating costs of a warehouse, e.g. for light, heating or cooling, electricity for technology, and labor costs. Other operating costs incur as a result of placing items into stock and release from stock, including related transportation costs.

Leakage (theft) of items from the warehouse is another possible cost driver. The most important cost aspect is, however, often due to the capital lockup of stored goods. These costs can be calculated similar to an interest rate for any other kind of investment.

Thus, on the one hand, there are various cost components which make warehousing costly and which suggest avoiding an inventory whenever possible. On the other hand, too small inventories are also undesirable. Stock-outs which occur when some goods are not or not in a sufficient quantity in stock, when customers want to buy them, cause opportunity costs, i.e. costs due to lost revenues. In most planning models it is assumed that customer demand should always be covered (whenever possible) although economic considerations may as well suggest not covering the demand when the profit contribution of the respective activities becomes negative.

For these reasons, there is a strong need for having good planning techniques for the inventory which minimize storage costs and, at the same time, make sure, that logistics needs are sufficiently fulfilled. Basic questions in inventory management concern the inventory levels and, along with that, the economic order quantities.

4.2 Economic Order Quantities and Safety Stocks

Two basic strategies are possible to fulfill the main logistic task which is to have the right item in the right quantity available at the right place: to store the goods in sufficient quantities where they are needed or to transport them to that place when they are needed. So, theoretically, warehousing could be avoided—at least partially—by using transportation. This idea has become popular under the slogan just-in-time (JiT) production where companies reduce their previously large warehouses of procured materials and agree with suppliers to deliver the required goods punctually, i.e. just before they are needed (see, e.g. Waters-Fuller 1995).

This concept has its origin in the car manufacturing industry, and it has been argued occasionally that it was enforced by the larger market power of car manufacturers compared to their component suppliers. This argument relies on the assumption that warehousing costs (and other costs) are shifted from the customers (car manufacturers) to their suppliers. Advocates of JiT, however, argue that a better cooperation between suppliers and customers can reduce the total quantities stored and thus the related warehousing costs of supplier and customer, who both can benefit from a win-win situation.

An obvious problem of the JiT concept is, however, that it requires frequent deliveries from the supplier. For the production on the same day, for instance, every morning a shipment of the necessary materials is required if no own stock of these materials exists. This may be adequate for car manufacturing where large amounts of materials are processed but for smaller amounts of materials the transportation costs of the frequent deliveries may consume the cost savings which result from reduced (or repealed) inventories.

This leads us to the question of how often deliveries of a material should take place, or vice versa, to the question of what an economic order quantity is. To find an answer, warehousing and transportation costs must be considered together: The smaller the ordered quantities, the smaller the inventory and related costs but the higher the transportation costs due to more frequent deliveries. The larger the ordered quantities, the smaller the transportation costs but the higher the warehousing costs.

Let us formulate a simple mathematical model to analyze this situation and to come to more definitive results: Let us assume that the consumption or demand for the inventory is steady-going, i.e. the number of items removed from the inventory per time unit is constant over time (no fluctuations of demand). Moreover, let us assume that the company orders always an equal amount of the considered material, denoted by x, and that new orders arrive just when the old inventory is used up. The development of the inventory can then be illustrated by Fig. 4.1

In such a situation, the average inventory is $x/2$, i.e. half of the ordered quantity.

Let us assume we want to plan the inventory for a planning horizon of T periods (e.g. T days) and that a total quantity of D is required during this planning horizon. The storage costs are assumed to be h per unit of the material and per period. The average storage costs per period are then $h \cdot x/2$ and the total storage costs in the planning horizon are $T \cdot h \cdot x/2$.

For the calculation of the transportation costs we make another roughly simplified assumption: It is assumed that each delivery causes fixed costs of u, i.e. the costs of delivery are independent of the delivered quantity. Since D items are required during the planning horizon, D/x deliveries are needed. This leads to transportation costs of $u \cdot D/x$. Note that the same model could also be applied for a planning of production (instead of external procurement) where u refers to setup costs of a production process.

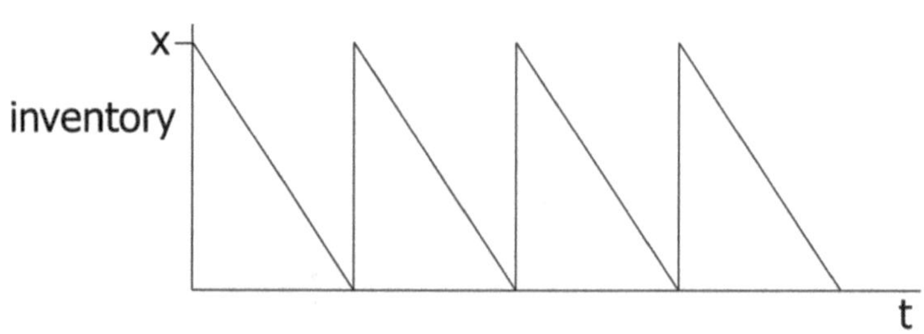

Fig. 4.1 Development of inventory in case of a steady-going consumption

The total costs of procurement are the total storage costs plus the total ordering costs, i.e. $u \cdot D/x + x/2 \cdot h \cdot T$. Note that the costs for the materials themselves are not considered as we assume that always in total the amount of D is ordered for a given price, irrespective of larger or smaller ordered quantities. The same consideration applied to the case of internal production neglects variable production costs which are assumed to be constant so that it is irrelevant for them whether smaller or larger lot-sizes are produced.

Because we want to minimize the total costs of procurement, we can simply check the first and second derivative of this cost function (with respect to the variable x) and thus obtain the following well-known square root formulae for x which fulfills the necessary and sufficient condition for a minimum (x_o is the economic order quantity, i.e. the value for x which leads to the lowest procurement costs):

$$x_o = \sqrt{2Du/hT} \tag{4.1}$$

In the literature, there are other versions of this formula, which are based on somewhat different formulations of the inventory model. For instance, we might express the inventory costs in the form of a holding cost rate per year, denoted as h^r, similar to an interest rate. Here it is assumed that costs due to the capital lockup are the dominant component of warehousing costs. The warehousing costs of a year can then be expressed by multiplying l with the average inventory multiplied with the purchase price of a unit of the material, denoted by p. The resulting formula for an economic order quantity is then:

$$x_o = \sqrt{2Du/ph^r} \tag{4.2}$$

where u are the fixed costs per delivery (as before) and D is the required quantity of the material per year.

The above inventory model can be summarized by the following three assumptions: constant inventory usage rates over time, storage costs proportional to the inventory levels, and procurement costs proportionally to the number of procurement operations (i.e. a constant cost for each procurement operation such as a set-up cost for a respective production or fixed transportation cost in the case of external procurement). This model dates back to the basic contributions by Harris (1913) and Andler (1929). It is also denoted as lot-sizing model with fixed set-up costs and constant unit inventory costs, as uncapacitated single item lot-sizing problem (USILSP), or as formulae for calculating the economic lot size (ELS) or the economic order quantity (EOQ).

It appears to be easy to use these results which can be expressed in a rather simple equation for the economic order quantities. Unfortunately, the assumptions of the model are not very realistic in most real-life cases so that it does not appear to be suitable. Some of the unrealistic assumptions could be remedied quite easily. For instance, it is quite simple to consider lead times for delivery, reorder points, or given levels of safety stocks (minimum stocks). In other cases, more difficult models and calculations are required.

Let us consider some of the easy extensions of the model: We assume a lead time of *LT* periods, i.e. it takes *LT* periods from ordering the material until its receipt. Let d be the (average) demand per period, i.e. $d = D/T$. Then the reorder point (or reorder level) *RP* is given by

$$RP = dLT, \tag{4.3}$$

i.e. when the inventory reaches this level a replenishment order for the material is required.

Another standard extension of the model is to consider safety stocks. A safety stock can be described as the minimum level of inventory because of fluctuations in demand or for other unforeseen circumstances. If such a safety stock level *SS* is to be applied, the reorder point is

$$RP = dLT + SS. \tag{4.4}$$

This formulation already leads beyond the simple inventory model which assumes a constant and thus certain demand per period. For most companies, their customer demand and also their internal demand is, however, neither constant nor certain.

In that case, the non-constant demand is the smaller problem. The assumption of an average inventory of $x/2$ may still be a good estimator for the actual situation and known fluctuations of demand may be well considered in calculating reorder points. The consideration of stochastic demand is more difficult. The most common model of accounting for this is as follows. It is assumed that the company uses some kind of prediction model for the future demand. This does not need to be very sophisticated. Even assuming a constant future demand would work. Assuming that the actual demand has a normal distribution, it is possible to calculate the prediction error for the demand

$$\sigma_d = \frac{1}{n}\sum_{t=1}^{n}\left(d_t - \hat{d}_t\right)^2 \tag{4.5}$$

where d_t is the actual demand in period t, $\hat{d}_t$ is the predicted demand in period t, and n is the number of considered periods from the past. In a similar way, the prediction error for the lead times can be determined (see, e.g., Chopra et al. (2004) for a more thorough discussion):

$$\sigma_{LT} = \frac{1}{m}\sum_{t=1}^{m}\left(LT_t - \overline{LT}\right)^2 \tag{4.6}$$

where LT_t is the lead time of the tth past delivery, $\overline{LT}$ is the average (or planned) lead time, and m is the number of considered past deliveries. The reorder point RP can then be calculated as follows:

$$RP = \overline{d}\overline{LT} + Z\sqrt{\overline{LT}\sigma_d^2 + \overline{d}^2\sigma_{LT}^2} \tag{4.7}$$

The parameters used in this equation are as follows:

$\overline{LT}$: average lead time
$\overline{d}$: average demand per period
Z : service level
σ_d: standard deviation of demand
σ_{LT}: standard deviation of lead time

Due to the harsh assumptions of this model, it is often not usable in practice. For instance, procurement costs may depend on the ordered quantities in the form of a constant or linear function with jump discontinuities. In such situations, optimal results cannot be simply calculated by a formula. Instead, more complex optimization models are to be derived which can be solved, for instance, by metaheuristics.

A more realistic but also more complex variant of the lot-sizing model is its dynamic version with assumed given time-varying demand for the forthcoming periods. This model has first been studied by Wagner and Whitin (1958) who formulate it in a dynamic programming version and propose a respective algorithm for its solution. For an easier formulation and also for taking care of variants of the dynamic lot-sizing problem, various specific heuristics have been developed such as the Silver and Meal (1973) heuristic.

4.3 Capacitated Lot-Sizing Problems

The lot-sizing problem becomes significantly more complex when additional constraints are to be taken into account, especially capacity constraints for the required resources. The single-item capacitated lot-sizing problem (SICLSP) can be defined as follows (see, e.g., Mateus et al. 2010):

$$\min\sum_{t=1}^{T} (c_t x_t + h_t i_t + u_t y_t) \tag{4.8}$$

$$i_{t-1} + x_t = d_t + i_t, t \in \{1, \ldots, T\} \tag{4.9}$$

$$a x_t + b y_t \leq w_t, t \in \{1, \ldots, T\} \tag{4.10}$$

$$ax_t \leq w_t y_t, t \in \{1,\ldots,T\} \tag{4.11}$$

$$y_t \in \{0,1\} \quad t \in \{1,\ldots,T\}, \tag{4.12}$$

$$x_t, i_t \in N, \quad t \in \{1,\ldots,T\} \tag{4.13}$$

Here, T is the planning horizon comprising the periods $\{1,\ldots,T\}$, c_t are the unit costs of production, h_t the unit costs (per period) of inventory, and u_t the setup costs for production. The decision variables are x_t, the produced amounts in period t, and y_t which is 1 if production takes place in t, and 0 otherwise. The inventory levels i_t are further decision variables but governed by the inventory balance equation (4.9) which specifies the relationship between previous inventory, production, demand, and resulting new inventory. Possibly, a starting inventory i_0 and a desired end inventory i_T are given. Further parameters given for the problem are the demand values d_t. The objective is the minimization of costs (4.8) which include production, holding, and setup costs. Equation (4.10) specifies the capacity constraint which assumes that the production of each unit requires a capacity (like time) of a while the setup of production requires a capacity of b which must be covered by the total available capacity w_t in each period t. Equation (4.11) is for making sure that y_t is 1 if $x_t > 0$, and 0 otherwise.

Further restrictions Eqs. (4.12) and (4.13) concern the domain of the variables which are Boolean values for y_t and natural numbers (nonnegative integers) for x_t and i_t. In particular, these values may not become negative. Alternatively, it could be assumed that the last variables take nonnegative real values.

Without Eq. (4.10) the problem is denoted as uncapacitated single item lot-sizing problem which can be solved in small polynomial time while the additional capacity constraint makes the problem NP-hard (Brahimi et al. 2006; Florian et al. 1980) and thus an interesting candidate for the application of metaheuristics.

Frequently, and because of its practical importance, multi-item versions of lot-sizing problems are considered. A generic extension of the above SICLSP Eqs. (4.8)–(4.13) is then the following multi-item capacitated lot-sizing problem (MICLSP):

$$\min \sum_{p=1}^{P} \sum_{t=1}^{T} (c_{pt}x_{pt} + h_{pt}i_{pt} + u_{pt}y_{pt}) \tag{4.14}$$

$$i_{p(t-1)} + x_{pt} = d_{pt} + i_{pt}, t \in \{1,\ldots,T\}, p \in \{1,\ldots,P\} \tag{4.15}$$

$$\sum_{p=1}^{P} (a_p x_{pt} + b_p y_{pt}) \leq w_t, t \in \{1, \ldots, T\} \tag{4.16}$$

$$a_p x_{pt} \leq w_t y_{pt}, t \in \{1,\ldots,T\}, p \in \{1,\ldots,P\} \tag{4.17}$$

$$y_{pt} \in \{0,1\} \quad t \in \{1,\ldots,T\}, p \in \{1,\ldots,P\} \tag{4.18}$$

$$x_{pt}, i_{pt} \in N, \quad t \in \{1,\ldots,T\}, p \in \{1,\ldots,P\} \tag{4.19}$$

Here it is assumed that there are P products with indexes in {1, . . ., P} which are to be planned during the planning horizon T. The objective function (4.14) is again the minimization of costs which sum up the production, holding, and setup costs for all products and all periods. The balance inventory equations (4.15) and the logical condition for the y variables (4.17) are now to be observed for all products. For the capacity constraint (4.16), capacity consumption due to production and setup is summed up for all products, i.e. we still assume that there is a single capacity constraint for all products. A similar version of the MICLSP with additional backordering can be found, e.g., in Mateus et al. (2010).

Note that the uncapacitated version, i.e. the above problem without Eq. (4.16) is still polynomial as it can be solved by solving an uncapacitated single item lot-sizing problem for each problem separately. The MICLSP (4.14)–(4.19) is, of course, NP-hard just like the SICLSP which is a special case of it.

Typically the items to be planned in a company are embedded in a hierarchical production structure, i.e. end items to be produced require components which have to be produced or procured as well. The relationships between items are described in the Bills of Materials (BOM) which usually assume that there is a fixed amount of various components necessary to produce one unit of the considered item (see Fig. 4.2 for an example). Usually, the resulting graph is assumed to be acyclic. (Special situations with cyclic production structure may occur in the chemical or food industry where we also find coupled production, i.e. several, sometimes undesired products resulting from a single production process.)

As a consequence of a hierarchical production structure, the demand for an item is not only the external demand (also denoted as primary demand) which is assumed to be given (e.g. by forecast models) but additionally the internal demand (secondary demand) which results from the fact that we plan to produce successor items in the production structure which require the considered item.

A simple variant of a respective planning problem is the multi-level uncapacitated lot-sizing problem (MLULSP, sometimes just denoted as multi-level lot-sizing problem, or MLLSP) which can be defined as follows: the production of P items is to be planned which are connected by a given production structure so that any external demand during T periods is satisfied. The objective Eq. (4.20)

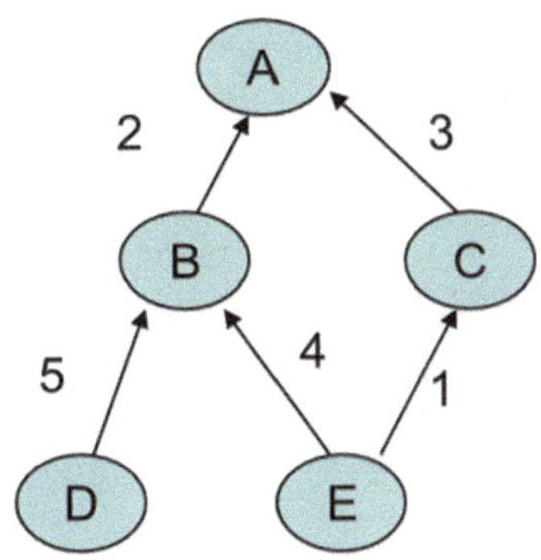

Fig. 4.2 Example of a Bill of Materials (BOM). The *numbers next to the arrows* indicate the respective production coefficients. For producing 1 piece of A, 2 pieces of B, 3 pieces of C, 10 pieces of D, and 11 pieces of E are required

here is to minimize the total costs which consist of inventory holding and setup costs (see, e.g., Homberger 2010).

$$\min\sum_{p=1}^{P}\sum_{t=1}^{T}(h_p i_{pt} + u_p y_{pt}) \tag{4.20}$$

$$i_{pt-1} + x_{pt} = d_{pt} + i_{pt}, p\in\{1,\ldots,P\}, t\in\{1,\ldots,T\} \tag{4.21}$$

$$d_{pt} = \sum_{j=1}^{P} v_{pj}x_{jt} + ed_{pt}, p\in\{1,\ldots,P\}, t\in\{1,\ldots,T\} \tag{4.22}$$

$$x_{pt} \leq My_{pt}, t\in\{1,\ldots,T\}, p\in\{1,\ldots,P\} \tag{4.23}$$

$$y_{pt}\in\{0, 1\} \tag{4.24}$$

$$x_{pt}, i_{pt}\in N, t\in\{1,\ldots,T\} \tag{4.25}$$

Compared with the MICLSP, the main difference is the calculation of total demand. It is assumed that some external demand ed_{pt} for item p in period t is given. v_{pj} are the production coefficients which define how many units of item p are required for producing one unit of item j. If some item j is not a successor of p then $v_{pj}=0$. The total demand for item p in period t is then given by Eq. (4.22) which adds the external and the internal demand given by the multiplication of successor production quantities with the production coefficients. Instead of summing up over all items, we could also just build the sum over all successors of item p.

Further differences to the MICLSP are that production costs are neglected which is possible if the production unit costs are identical in each period for every product. (A version of the problem including the production costs is, for instance, treated in Dellaert et al. (2000).) The missing capacity constraint leads to a reformulation of Eq. (4.23) compared to Eq. (4.17) where the capacity limit is replaced by a sufficiently large number M. Often the integer assumptions in Eq. (4.25) are left out while still assuming that x_{pt} and i_{pt} are nonnegative.

As shown by Arkin et al. (1989), the MLULSP is NP-hard for production structures where each item can have more than one successor and predecessor. The introduction of capacity constraints is a further, practically more relevant version of the problem and also an extension of the MICLSP. This problem is sometimes denoted as general capacitated lot-sizing problem (GCLSP) but there are several variants of this problem with additional constraints and other modifications, e.g. the consideration of overtime capacities and costs (see, e.g., Xie and Dong 2002).

A further extension of the MICLSP is the coordinated capacitated lot-sizing problem (CCLSP) which additionally considers individual and joint setup costs when production in a period occurs (Robinson et al. 2009). While the CCLSP is NP-hard, this also holds for its capacity-relaxed version, the coordinated uncapacitated lot-sizing problem (CULSP) (Arkin et al. 1989).

4.4 Solution Approaches for Capacitated Lot-Sizing Problems

As we have seen, capacitated lot-sizing problems in general but also some uncapacitated variants are NP-hard and, therefore, metaheuristics are interesting methods to solve them. Of course, there are also many contributions which focus on exact approaches such as branch & bound, branch & cut, and Lagrangian relaxation or which use rather simple problem-dependent heuristics. Further complications result from considering a multi-item problem situation (possibly with shared resources among the different items) or with considering multi-stage situations (as in Fig. 4.2). Apart from the complexity issue, these problems can become quite large because the number of variables is a product of the number of items (including predecessors in the production structure or materials to be procured) and the number of planning periods. It should be noted that lot-sizing problems exist in various further variants, some of them linking the problem with other operational problems such as scheduling (Karimi et al. 2003).

For the lot-sizing problems discussed above, the representation in a metaheuristic is a crucial decision for the success. For the decision variables, it is often a straightforward and efficient way to represent them by corresponding data structures, e.g. quantities like x_{pt} as integer numbers if they are assumed to be integer, as floating point number if they are assumed to be real values, or for binary variables like y_{pt} to use respective binary variables in the used programming language. Since there is some redundancy among the variables (y_{pt} is 1 if $x_{pt} > 0$ and 0 otherwise) we might consider skipping the y variables. Then, however, we should consider how it becomes sufficiently probable to arrive at 0 values for the x_{pt} variable. For instance, if we change these variables during the algorithm by random real values (as, e.g., in various evolutionary algorithms), arriving exactly at a new value of 0 is rather unlikely. If we keep the y variables, we, of course, need to make sure that such conditions like Eqs. (4.17) or (4.23) are kept. This could be achieved by adjusting between the x_{pt} and the y_{pt} values. For instance, first the y_{pt} variables are determined. If $y_{pt} = 0$ then x_{pt} is set to 0 as well. Otherwise, it is necessary to make sure that x_{pt} is set to a positive value.

For the observation of the other constraints, various approaches are, in general, suitable. Apart from generic approaches like recalculation of variable values or the usage of a penalty function approach, specific repair operations have often shown to be superior. For the lot-sizing problems there are some quite obvious approaches. If capacity constraints such as Eq. (4.16) are being violated, it might be a simple approach to reduce the related production amount x_{pt} for one or several products (possibly let random factors decide which products to reduce). Inventory balance equations like (4.15) can mainly be used for updating the inventory levels from period to period when d_{pt} are given and x_{pt} are set. A possibly resulting negative inventory in some period could be "repaired" by increasing the production in the respective period. If this is not possible (because of capacity constraints) or desired (because $y_{pt} = 0$ and should not be changed, if possible), the production in previous periods could be increased. Of course, due to further exhausted capacities in the previous periods (or when arriving at the beginning of the planning horizon), this might not be possible either. For a capitated version of the lot-sizing problems it is always possible that no feasible solution exists, i.e. demand cannot be fully satisfied in each period. Therefore, it might be a good idea to consider unsatisfied demand as a penalty term for the objective function. Occasionally, we find this idea already in the respective mathematical problem formulation. Unsatisfied demand is then usually expressed as shortage costs which become a component of the objective function (see, e.g., Absi and Kedad-Sidhoum 2008). Another possibility to cope with the constraints is demonstrated in Haase and Kohlmorgen (1996) who use a (real-valued) random key encoding for representing solutions in the framework of a genetic algorithm. The coefficients of the random key vector help to determine the start of production activities in line with capacity constraints and to meet the assumed demand.

With respect to metaheuristics used for solving lot-sizing problems the following methods can be found:

- genetic algorithms (Akrami et al. 2006; Dellaert et al. 2000; Gutierrez et al. 2001; Gaafar 2006; Homberger 2008; Kämpf and Köchel 2006; Haase and Kohlmorgen 1996; Woarawichai et al. 2011)
- tabu search (Akrami et al. 2006; Berretta et al. 2005; Gopalakrishnan et al. 2001; Xie and Dong 2002)
- neighborhood search (Almada-Lobo and James 2010; Clark 2000; Muller et al. 2012)
- memetic algorithms (Berretta and Rodrigues 2004; Torabi et al. 2006)
- hybrid metaheuristics (Berretta et al. 2005; Staggemeier et al. 2002; Mishra et al. 2011)
- particle swarm optimization (Dye and Hsieh 2013)
- simulated annealing (Homberger 2010)
- GRASP with path-relinking (Nascimento et al. 2010),
- ant colony optimization (Pitakaso et al. 2006)
- a cross entropy-based metaheuristic (Caserta and Rico 2009)

Let us note that there are also problem-specific designed heuristics such as in França et al. (1997). Recent reviews of mathematical and metaheuristic approaches are given by Jans and Degraeve (2007), Buschkühl et al. (2010) and Goren et al. (2010). The last review focuses on genetic algorithms as one of the most widely used methods for complex lot-sizing problems.

Based on published results, it is not easy to judge how metaheuristics perform in comparison. In many of the studies, comparisons of methods are not conducted. Another difficulty is that for most of the considered problem types there are no generally accepted collections of test problems which might search for benchmarking the different suggested methods. More recent publications show, that meanwhile test problems are often used repeatedly and begin to become established as benchmark instances.

In Akrami et al. (2006) both a genetic algorithm and tabu search show clearly better results on randomly generated test instances of a combined lot-sizing and scheduling problem in comparison with an "optimal enumeration method". In Almada-Lobo and James (2010) variable neighborhood search and tabu search are compared with a standard optimization software (CPLEX) based on a branch & bound method. For larger, randomly generated test instances, both metaheuristics outperform the standard method with tabu search remaining mostly superior for the considered problem sizes. Using a series of test problems (with different characteristics) in Berretta and Rodrigues (2004) a memetic algorithm is compared with optimal CPLEX solutions, with the problem-specific heuristics in França et al. (1997) and with those of another Lagrangean-based heuristics. In general, the result showed that the memetic algorithm was able to calculate better results than the compared previous heuristics. Similar results for a hybrid metaheuristics were shown in Berretta et al. (2005) and for genetic algorithms in Staggemeier et al. (2002) and Woarawichai et al. (2011). Dellaert et al. (2000) achieve good results for a GA compared with two simple heuristics and a time-constrained branch & bound-implementation. Moreover, superior results in comparison with a modified Silver-Meal heuristic are shown in Gaafar (2006). Further improvements of genetic algorithm are studied in Homberger (2008) and compared with other GA variants and an ant colony system approach. Before that, this ant colony system approach (Pitakaso et al. 2006) has shown superior results compared to some other heuristic approaches in the literature. In Homberger (2008) the new GA performs very well and is able to calculate some new best solutions for some established test instances. Further competitive results under a similar evaluation framework are obtained for simulated annealing in Homberger (2010). In Nascimento et al. (2010) computational experiments with a GRASP with path relinking heuristic lead to some new best solutions compared with those from the literature for the multi-plant capacitated lot-sizing problem.

4.5 Planning Warehouse Operations

Apart from answering the question how much should be produced and kept in stock, the planning of specific operations relating to warehousing and inventory management received considerable attention in the academic literature. One of the respective activities is order picking. (Transfers of materials to stock involve similar problems but are usually less time-critical and therefore not so often considered in optimization approaches.) Basically, order picking can be assumed as a specific type of transport combined with some additional activities. For instance, a worker walks through the warehouse to specific storage locations with items required for an order, picks them, and delivers them to a predefined location (see Fig. 4.3).

This type of problem is a typical round trip problem like the capacitated vehicle routing problem and can be solved by similar techniques (see above). Capacity limits are caused by the physical limitations when transporting loads or result from

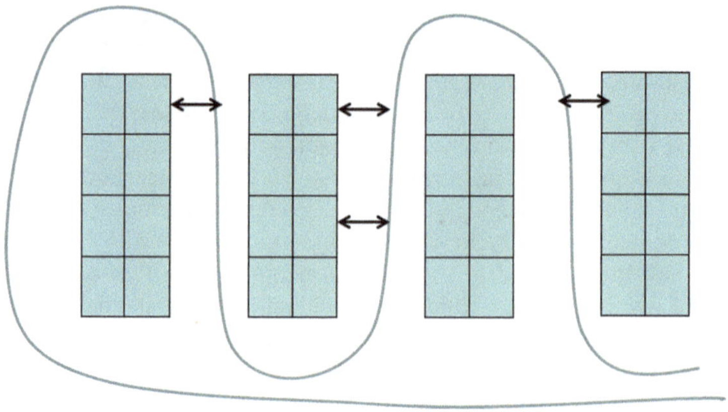

Fig. 4.3 Example of a round trip through warehouse aisles with four pickups of items

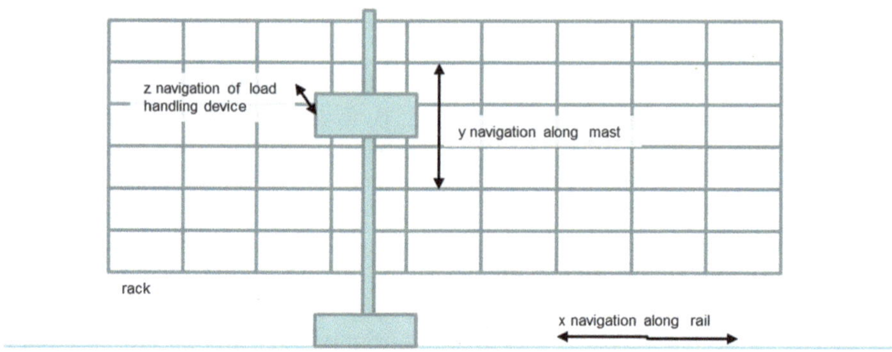

Fig. 4.4 Schematic design of a storage/retrieval system (AS/RS) operating a rack

capacity constraints depending on the used devices (e.g. a bin or a trolley). Due to the capacity constraints, large orders must possibly be split while smaller orders can be combined and executed in a single warehouse round trip (if this is supported from an organizational perspective). There are, however, modifications from standard CVRPs. Firstly, a required item may be available at several storage locations which allows for further possibilities of optimization. Secondly, the sequence of order fulfillment matters, especially when orders are processed simultaneously or in an overlapping fashion.

In warehouses, automated systems such as automated storage/retrieval systems (AS/RS) are often used for order picking or related activities like supplying "goods-to-man" order picking stations. Typically, an AS/RS is a computer controlled device operating along a track in the aisle between two racks. It is equipped with a mast which covers the height of the racks and along which a shuttle moves vertically to access defined storage locations for picking up or dropping off loads into the rack (see Fig. 4.4). Modern AS/RS often have a capacity larger than 1 so that several loads can be transported simultaneously. This also allows to formulate their operations as round trip problems. Sometimes, they do not have only a single point for interacting with other logistic systems so that more freedom concerning starting and end locations in a trip can be considered. Again, the sequencing of single transport orders and their interleaving is important for optimizing operations. Especially, the combination of movements for transfer to stock and for release from stock helps to avoid empty drives.

4.6 Storage Locations

Another aspect which significantly influences the efficiency of order picking and other activities is the specific determination of storage locations, also known as storage location assignment problem. For determining a suitable location in a warehouse, traditionally physical properties of the goods were considered as well as commodity groups. Storing similar products close to each other makes sense, because it is easier for the customer to find the desired products. This concept still dominates in the retail business where customers pick up their purchases. In warehouses where the order picking is done by qualified personal, today, the task of finding the products is usually supported by providing information about the specific locations so that a dedicated assignment of storage locations is no longer required. Instead, storage locations can be assigned in an arbitrary, random-like fashion as long as the storage locations fulfill the physical requirements of the goods (size, weight, etc.). Of course, a purely random assignment of locations is not sensible. Instead, it makes sense to assign locations so that the respective activities like order picking become more efficient. A simple rule could be to store often needed articles at locations which are easily (quickly) accessible while less frequently required goods could be stored at less advantageous places. Also, in such a consideration the required space per item matters.

First of all, there are many specific heuristics for solving such a problem, see, e.g., Goetschalckx and Ratliff (1990). The most prominent among these heuristics and rules is the cube-per-order index (COI) heuristic by Heskett (1963). The COI is the ratio of the volume of an item (space requirement in storage) to its demand (turnover per period). Products with small COI should be located closer to the shipping area and those with larger COI further away.

Better results can be expected from solving the storage location assignment as an optimization problem (stock location assignment problem, SLAP). In the literature two basic approaches are investigated: The individual assignment of locations for goods and the assignment of goods to classes of similar goods (e.g. similar demand or similar COI) and determining the space for these groups (class-based storage location assignment).

With respect to metaheuristics, the problem is solved, for instance, by genetic algorithms (Park and Seo 2010), by a Pareto and Niche Genetic Algorithm (Li et al. 2008), by simulated annealing (Muppani and Adil 2008), and by hybrid metaheuristics such as a combination of Particle Swarm Optimization and Artificial Bee Colonies (Hu et al. 2012).

The storage location assignment problem can also be combined with other aspects of optimizing warehouse operations, e.g. taking care of an efficient interleaving of AS/RS movements, e.g. in Chen et al. (2011) where this problem is tackled by a tabu search approach.

Another aspect in assigning storage locations is the need to reassign locations due to seasonal fluctuations and other changes in demand or product mix. In Kofler et al. (2011) it is shown that such reorganizations can be carried out periodically without too much operational effort.

Similar to transport problems, inventory-related optimization problems exist in many other variants, some of them focusing on specific details (e.g. warehouse operations aspects), other linking them with subordinate aspects such as related transportation processes (also denoted as inventory routing) or considering multiple locations of inventory and other supply chain aspects.

4.7 Inventory Routing

Most of the previous problems in inventory management more or less neglect the fact that warehouse operations are usually closely linked-up with transport activities. For instance, the fulfilment of customer orders first requires the availability of the needed items (or the ability to produce or procure them) and secondly their transport to the customers. Inventory routing problems try to combine both activities in a single planning and optimization model.

The easiest case is the model described for the EOQ which considers transport and holding costs in a most simple way (see Sect. 4.2). But already slightly different situations while still having a single supply and still dealing with a single customer can be much more complicated. For instance, if transports are made by vehicles with limited capacities and transport costs depend only on the number of vehicle tours, e.g. a fixed transport cost per vehicle, the situation may change and an optimal solution just by calculating the derivative of an objective function (as assumed for the EOQ formulae) does no longer lead to the optimal solution.

The multicommodity version of this type of problem is denoted as Single Link Shipping Problem (SLSP) (see Bertazzi and Speranza 2012). Here a supplier has to ship a set of P products to a customer with a constant demand $d_p, p \in \{1,\ldots, P\}$. For the supplier it is assumed that products are "generated" with the same constant amount per period d_p. The products are to be shipped with capacity-constrained vehicles and the transportation costs are assumed to be c for each trip (independent of the used capacity of the vehicle). In addition to the transport costs, inventory costs arise assuming that for each item a holding cost h_p is given regardless of whether it is held at the source or at the destination. The variables to be determined are the shipped quantities x_{pt} for item p in period t and, depending on that, the number of used vehicles, y_t. The objective is to minimize the total costs while fulfilling the entire demand in all periods. Mathematically the problem can be formulated as follows (modified from Bertazzi and Speranza 2002; also cf. Speranza and Ukovich 1994):

$$\min\sum_{t=1}^{T}\sum_{p=1}^{P}h_p\left(i_{pt}^A+i_{pt}^B\right)+\sum_{t=1}^{T}cy_t \tag{4.26}$$

$$\sum_{t=1}^{T}x_{pt}=d_pT,\quad p\in\{1,\ldots,P\} \tag{4.27}$$

$$\sum_{p=1}^{P}a_px_{pt}\le qy_t,\quad t\in\{1,\ldots,T\} \tag{4.28}$$

$$i_{pt}^A=i_{p0}^A+d_pt-\sum_{t'=1}^{t}x_{pt'},\quad t\in\{1,\ldots,T\},p\in\{1,\ldots,P\} \tag{4.29}$$

$$i_{pt}^B=i_{p0}^B+\sum_{t'=1}^{t}x_{pt'}-d_pt_t,\quad t\in\{1,\ldots,T\},p\in\{1,\ldots,P\} \tag{4.30}$$

$$i_{pt}^A\ge 0,\quad t\in\{1,\ldots,T\},p\in\{1,\ldots,P\} \tag{4.31}$$

$$i_{pt}^B-x_{pt}\ge 0,\quad t\in\{1,\ldots,T\},p\in\{1,\ldots,P\} \tag{4.32}$$

The objective function in Eq. (4.26) describes the minimization of the sum of holding costs and transport costs. Here the total inventory of each product is equal to the inventory in the source location plus the inventory in the destination location, i.e.

$$i_{pt}=i_{pt}^A+i_p^Bt,p\in\{1,\ldots,P\},t\in\{1,\ldots,T\} \tag{4.33}$$

Equation (4.27) makes sure that the sum of all shipments covers the total amount of products available during the considered time horizon. Equation (4.28) is for keeping the capacity constraints of the used vehicles: a_p is the capacity requirement (volume) per unit of product p while q is the offered capacity of a vehicle. The inventory at source location A is calculated in Eq. (4.29) as the sum of starting inventory and amounts becoming available until period t minus the shipped quantities until t. In a similar way, Eq. (4.30) describes the inventory at the destination location B as starting inventory plus arriving shipments minus demand until considered period t. Equations (4.31) and (4.32) make sure that the inventory in source and destination location cannot become negative. Note that the arriving shipment is subtracted from the inventory in Eq. (4.32) because the respective variable describes the value after the arrival.

In Bertazzi and Speranza (2002) several variants of the single link problem are considered, especially a time-continuous version and a time-discrete version with assumed constant frequencies of shipments.

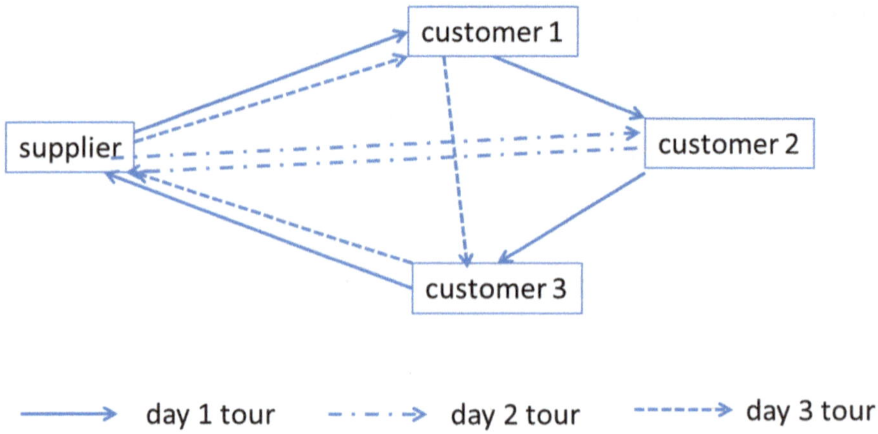

Fig. 4.5 Example of an inventory routing problem with one supplier and three customers leading to three different routes on different days

Of course, this single-link case is not particularly realistic. For that reason, mostly a situation is assumed which considers the delivery to several customers (Fig. 4.5).

Such an inventory routing problem can briefly be described as follows (see Archetti et al. 2007): It is assumed that there is a set of customers (e.g. retailers) $\{1,\ldots,N\}$ to be supplied by a single supplier over a time horizon $\{1,\ldots,T\}$. At the supplier a quantity d_{0t} is produced (or procured) at each time t, and at customer n a quantity d_{nt} is required (and depleted) at each time t. The inventory level at each customer is denoted as i_{nt} and we assume that initial inventory levels i_{n0} are given. Moreover, it is assumed that maximum inventory levels denoted as U_i are to be observed at any customer i. The inventory levels at the supplier are denoted by i_t starting with a given inventory level of i_0. The unit inventory costs at the supplier are denoted by h_0 whereas h_n are the unit inventory costs for customer n.

Goods are sent from the supplier to the customers by a vehicle with capacity q. Customers can be served in any arbitrary sequence by building an appropriate route. The transportation costs are denoted as c_{ij} for locations $i,j \in \{0,\ldots,n\}$ denoting the location of the supplier (in case of index value 0) or a location of the customers (in case of index value ≥ 0). The mathematical formulation of the problem is then as follows:

$$\min \sum_{t=1}^{T+1} h_0 i_t + \sum_{n=1}^{N} \sum_{t=1}^{T+1} h_n i_{nt} + \sum_{i=0}^{N} \sum_{j=0}^{i-1} \sum_{t=1}^{T} c_{ij} y_{ij}^t \tag{4.34}$$

$$i_t = {}^t i_{t-1} + {}^{t-1t-1} d_{0t-1} - {}^{0t-10t-1} \sum_{i=1}^{N} x_{it-1}, \quad t \in \{1, \ldots, T+1\} \tag{4.35}$$

$$i_t \geq \sum_{i=1}^{N} x_{it}, \quad t \in \{1, \ldots, T\} \tag{4.36}$$

$$i_{it} = i_{it-1} + x_{it-1} - d_{it-1}, \ i \in \{1, \ldots, N\}, t \in \{1, \ldots, T+1\} \tag{4.37}$$

$$i_{it} \geq 0, \quad i \in \{1, \ldots, N\}, t \in \{1, \ldots, T+1\} \tag{4.38}$$

$$\sum_{i=1}^{N} x_{it} \leq q, \quad t \in \{1, \ldots, T\} \tag{4.39}$$

$$\sum_{i=1}^{N} x_{it} \leq q z_{0t}, \quad t \in \{1, \ldots, T\} \tag{4.40}$$

$$\sum_{j=1}^{i-1} y_{ij}^{t} + \sum_{j=i+1}^{N} y_{ji}^{t} = 2 z_{it}, \quad i \in \{1, \ldots, N\}, t \in \{1, \ldots, T\} \tag{4.41}$$

$$x_{it} \geq 0, \quad i \in \{1, \ldots, N\}, t \in \{1, \ldots, T\} \tag{4.42}$$

$$y_{ij}^{t} \in \{0, 1\}, \quad i, j \in \{1, \ldots, N\}, t \in \{1, \ldots, T\} \tag{4.43}$$

$$z_{it} \in \{0, 1\}, \quad i \in \{0, \ldots, N\}, t \in \{1, \ldots, T\} \tag{4.44}$$

The objective function (4.34) consists of the holding costs at the supplier, the holding costs of all customers, and the transport costs. Note that the sums of holding costs include period $T+1$ to consider inventory at the end of the planning horizon. The calculation of transport costs involves binary variables y^t_{ij} which define whether at time t a route with transport from node i to node j is involved. Also note that in this problem formulation undirected arcs are considered.

Equations (4.35) and (4.37) are the inventory balance equations for the supplier and the customers. Equation (4.36) is for preventing stock-outs at the supplier whereas Eq. (4.38) is for preventing negative inventory at the customers.

Equation (4.39) makes sure that the vehicle capacity is observed whereas Eq. (4.40) relates it to the binary variables z_{it} which define whether at time t node i is visited ($z_{it} = 1$) or not ($z_{it} = 0$).

Equation (4.41) is for specifying consistent tours which requires that for each visited customer (and for the supplier as well) there are two edges in the tour, one for driving to the customers, another for driving away from them.

Often there are additional constraints e.g. for a so-called order-up-to level inventory policy which says that if a customer is supplied by time t then the resulting inventory should reach a given level:

$$x_{it} = (U_i - i_{it}) z_{it}, i \in \{1, \ldots, N\}, t \in \{1, \ldots, T\} \tag{4.45}$$

Further constraints may be useful for an efficient modelling of the problem in a respective standard software, for instance a subtour elimination constraint (Archetti et al. 2007).

As one can easily imagine, apart from those standard formulations of inventory routing problems many variants are possible to represent practical situations in a better way. One possibility to distinguish different types of inventory routing problems is the consideration of time. Time can either be discrete, i.e. only periods of time such as days are considered; or time is measured in a continuous fashion. Another aspect concerns the time horizon: There are single-period models, multi-period models, and infinite horizon models (see, Moin and Salhi 2007).

Further distinctions concern the network structure. Besides the above cases with a single link and a single supplier but several customers, there can be, of course, also situations with multiple suppliers and customers. Alternative models for the deliveries are direct deliveries only (like in the single link case), routes with several customers and possibly deliveries without a single, central depot, i.e. with possible inter-customer deliveries. The latter case is in the literature (see, e.g. Coelho et al. 2013) also denoted as continuous case or as model with transshipments.

Very often, additional aspects are to be taken into account when solving inventory routing problems. Especially, it is often desired that the delivery plans show cyclical patterns or defined frequencies of delivery, e.g. deliveries every 3 days. Other requirements could be maximum inventory levels at customers, full load shipments only, prohibition for specific routes, e.g. direct delivery only, or the partitioning of customers to be served in a route (see, Bertazzi and Speranza (2012) for more examples on such constraints.)

Further additional constraints may relate to the inventory replenishment policy: In particular, maximum levels at the destinations need to be observed or there is an order-up-to level inventory policy (see above). Another possible policy is the Zero Inventory Ordering (ZIO) which allows to replenish stocks at a customer's only if their inventory level is down to zero.

Insufficient deliveries (which should not result in negative stock) may be considered as lost sales or may lead to back orders (i.e. deliveries should be caught up as soon as possible). Finally the transport means could be considered in a more differentiated way, so there could be a single vehicle, a defined number of vehicles or an unconstrained number of vehicles, the latter possibly with the goal to use as few vehicles as possible. If several vehicles are considered, they could be identical or heterogeneous with respect to capacity. Apart from observing the load capacity of vehicles a full load shipment policy may be applied which allows to start vehicle tours only with fully utilized capacities.

With respect to solution approaches, despite the high computational complexity the literature is still dominated by traditional mathematical optimization approaches on the one hand, and simple or problem-specific approaches on the other hand which are occasionally combined. For instance, in Campbell and Savelsbergh (2004) the problem is decomposed into two subproblems where one of them (the determination which customers to serve on which day and by which quantity) is solved by a branch and bound approach whereas the second phase (determination of routes) is solved by an insertion heuristic. Another example of a two-phase approach is Kang and Kim (2010) where specific heuristics are used in both phases.

An example of a two-phase approach involving a metaheuristics in the second phase is Sindhuchao et al. (2005). In the second phase they use a very large-scale neighborhood search (VLSN) approach whereas the first phase was solved by a column generation method.

García et al. (2013) use two metaheuristics to solve a multiobjective inventory routing problem: variable neighborhood search (VNS) and the evolutionary algorithm NSGA-II. The considered objectives are the transportation costs and the inventory costs. Using some real-life problem data, the VNS metaheuristic obtained better results. Vidović et al. (2014) suggest a variable neighborhood descent (VND) approach based on a constructive heuristic. The approach uses two types of VND search, first local intra-period search by improving the routes (using inter-route exchanges of destinations, removal and reinsertion of locations inside one route, and the exchange of route segments), then a large inter-period neighborhood search which is based on the idea of removing the delivery of a location from a route and shifting the delivered quantity to a different day. Compared with a mixed-integer problem formulation using the CPLEX software, VND leads for small problem sizes to results with costs less than 4 % higher but significantly shorter computation times whereas larger problem instances could only be solved by VND.

In Liu and Chen (2011) tabu search adopting different neighborhood search approaches is used for solving an extended inventory routing problem. In Lee et al. (2006) tabu search is applied as well for solving inventory routing problems. Numerical experiments show that the obtained solutions are not worse more than 4 % compared to optimal solutions while requiring reasonable computation times.

Another tabu search solution approach is investigated in Toptal et al. (2013). In various experiments this approach obtained optimal or near optimal solutions with a 0.31 % deviation from the respective lower bounds.

Tabu search is also used by Zhao et al. (2007) where it serves for determining the partitioning of the customers for building routes. A similar approach is studied in Zhao et al. (2008) where the partitions are obtained by a variable neighborhood search (VNS) method.

Occasionally, also specific CI approaches were used to take care of uncertainties or incomplete information during the modeling process of inventory routing problems. For instance, ant colony optimization is applied to a multi-item inventory routing problem with uncertain demand in Huang and Lin (2010). For the test runs, demand is here calculated as a normally distributed variable with known parameters. Another example is a fuzzy model of a multi-echelon inventory situation in a supply chain studied in Giannoccaro et al. (2003). Here a fuzzy approach serves to model uncertainties in both the demand and the assumed holding and backorder costs. An approach for inventory management based on a Markov decision processes model and using reinforcement learning is studied in Giannoccaro and Pontrandolfo (2002).

In general, however, it can be observed that fuzzy but also more traditional stochastic models have been neglected during research on inventory routing problems. A short review of this area can be found in Bertazzi et al. (2013).

References

Absi, N., & Kedad-Sidhoum, S. (2008). The multi-item capacitated lot-sizing problem with setup times and shortage costs. *European Journal of Operational Research, 185*(3), 1351–1374.

Akrami, B., Karimi, B., & Hosseini, S. M. (2006). Two metaheuristic methods for the common cycle economic lot sizing and scheduling in flexible flow shops with limited intermediate buffers: The finite horizon case. *Applied Mathematics and Computation, 183*(1), 634–645.

Almada-Lobo, B., & James, R. J. (2010). Neighbourhood search meta-heuristics for capacitated lot-sizing with sequence-dependent setups. *International Journal of Production Research, 48*(3), 861–878.

Andler, K. (1929). *Rationalisierung der Fabrikation und optimale Losgröße*. München: Oldenbourg.

Archetti, C., Bertazzi, L., Laporte, G., & Speranza, M. G. (2007). A branch-and-cut algorithm for a vendor-managed inventory-routing problem. *Transportation Science, 41*(3), 382–391.

Arkin, E., Joneja, D., & Roundy, R. (1989). Computational complexity of uncapacitated multi-echelon production planning problems. *Operations Research Letters, 8*(2), 61–66.

Berretta, R., França, P. M., & Armentano, V. A. (2005). Metaheuristic approaches for the multilevel resource-constrained lot-sizing problem with setup and lead times. *Asia-Pacific Journal of Operational Research, 22*(02), 261–286.

Berretta, R., & Rodrigues, L. F. (2004). A memetic algorithm for a multistage capacitated lot-sizing problem. *International Journal of Production Economics, 87*(1), 67–81.

Bertazzi, L., Bosco, A., Guerriero, F., & Lagana, D. (2013). A stochastic inventory routing problem with stock-out. *Transportation Research Part C: Emerging Technologies, 27*, 89–107.

Bertazzi, L., & Speranza, M. G. (2002). Continuous and discrete shipping strategies for the single link problem. *Transportation Science, 36*(3), 314–325.

Bertazzi, L., & Speranza, M. G. (2012). Inventory routing problems: An introduction. *EURO Journal on Transportation and Logistics, 1*(4), 307–326.

Brahimi, N., Dauzere-Peres, S., Najid, N. M., & Nordli, A. (2006). Single item lot sizing problems. *European Journal of Operational Research, 168*(1), 1–16.

Buschkühl, L., Sahling, F., Helber, S., & Tempelmeier, H. (2010). Dynamic capacitated lot-sizing problems: A classification and review of solution approaches. *OR Spectrum, 32*(2), 231–261.

Campbell, A. M., & Savelsbergh, M. W. (2004). A decomposition approach for the inventory-routing problem. *Transportation Science, 38*(4), 488–502.

Caserta, M., & Rico, E. Q. (2009). A cross entropy-Lagrangean hybrid algorithm for the multi-item capacitated lot-sizing problem with setup times. *Computers & Operations Research, 36* (2), 530–548.

Chen, L., Langevin, A., & Riopel, D. (2011). A tabu search algorithm for the relocation problem in a warehousing system. *International Journal of Production Economics, 129*(1), 147–156.

Chopra, S., Reinhardt, G., & Dada, M. (2004). The effect of lead time uncertainty on safety stocks. *Decision Sciences, 35*(1), 1–24.

Clark, A. R. (2000). A local search approach to lot sequencing and sizing. In M. Tucci & M. Garetti (Eds.), *Proceedings of the Third International Workshop of the IFIP WG 5.7 Special Interest Group on "Advanced techniques in production planning & control"* (pp. 145–152). Firenze: Firenze University Press.

Coelho, L. C., Cordeau, J. F., & Laporte, G. (2013). Thirty years of inventory routing. *Transportation Science, 48*(1), 1–19.

Dellaert, N., Jeunet, J., & Jonard, N. (2000). A genetic algorithm to solve the general multi-level lot-sizing problem with time-varying costs. *International Journal of Production Economics, 68* (3), 241–257.

Dye, C. Y., & Hsieh, T. P. (2013). A particle swarm optimization for solving lot-sizing problem with fluctuating demand and preservation technology cost under trade credit. *Journal of Global Optimization, 55*(3), 655–679.

Florian, M., Lenstra, J. K., & Rinnooy Kan, A. H. G. (1980). Deterministic production planning: Algorithms and complexity. *Management Science, 26*(7), 669–679.

França, P. M., Armentano, V. A., Berretta, R. E., & Clark, A. R. (1997). A heuristic method for lot-sizing in multi-stage systems. *Computers & Operations Research, 24*(9), 861–874.

Gaafar, L. (2006). Applying genetic algorithms to dynamic lot sizing with batch ordering. *Computers & Industrial Engineering, 51*(3), 433–444.

García, I., Pacheco, J., & Alvarez, A. (2013). Optimizing routes and stock. *Journal of Heuristics, 19*(2), 157–177.

Giannoccaro, I., & Pontrandolfo, P. (2002). Inventory management in supply chains: A reinforcement learning approach. *International Journal of Production Economics, 78*(2), 153–161.

Giannoccaro, I., Pontrandolfo, P., & Scozzi, B. (2003). A fuzzy echelon approach for inventory management in supply chains. *European Journal of Operational Research, 149*(1), 185–196.

Goetschalckx, M., & Ratliff, H. D. (1990). Shared storage policies based on the duration stay of unit loads. *Management Science, 36*(9), 1120–1132.

Gopalakrishnan, M., Ding, K., Bourjolly, J. M., & Mohan, S. (2001). A tabu-search heuristic for the capacitated lot-sizing problem with set-up carryover. *Management Science, 47*(6), 851–863.

Goren, H. G., Tunali, S., & Jans, R. (2010). A review of applications of genetic algorithms in lot sizing. *Journal of Intelligent Manufacturing, 21*(4), 575–590.

Gutierrez, E., Hernandez, W., & Süer, G. A. (2001). Genetic algorithms in capacitated lot sizing decisions. In *Computing Research Conference,* March 31, 2001, Puerto Rico. Accessed March 12, 2016, from http://www.ece.uprm.edu/crc/crc2001/papers/eliecer_gutierrez.PDF

Haase, K., & Kohlmorgen, U. (1996). Parallel genetic algorithm for the capacitated lot-sizing problem. In P. Kleinschmidt, A. Bachem, U. Derigs, D. Fischer, U. Leopold-Wildburger & R. Möhring (Eds.), *Operations Research Proceedings 1995* (pp. 370–375). Berlin/Heidelberg: Springer. Accessed March 12, 2016, from http://citeseerx.ist.psu.edu/viewdoc/download?doi=10.1.1.45.4394&rep=rep1&type=pdf

Harris, F. W. (1913). How many parts to make at once. *Factory, The Magazine of Management, 10* (2), 135–136, 152, reprinted in *Operations Research, 38*(6), 947–950.

Heskett, J. L. (1963). Cube-per-order index-a key to warehouse stock location. *Transportation and Distribution Management, 3*(1), 27–31.

Homberger, J. (2008). A parallel genetic algorithm for the multilevel unconstrained lot-sizing problem. *INFORMS Journal on Computing, 20*(1), 124–132.

Homberger, J. (2010). Decentralized multi-level uncapacitated lot-sizing by automated negotiation. *4OR: A Quarterly Journal of Operations Research, 8*(2), 155–180.

Hu, W., Wang, Y., & Zheng, J. (2012). Research on warehouse allocation problem based on the Artificial Bee Colony inspired particle swarm optimization (ABC-PSO) algorithm. *5th International Symposium in Computational Intelligence and Design (ISCID)* Vol. 1, (pp. 173–176). Piscataway: IEEE.

Huang, S. H., & Lin, P. C. (2010). A modified ant colony optimization algorithm for multi-item inventory routing problems with demand uncertainty. *Transportation Research Part E: Logistics and Transportation Review, 46*(5), 598–611.

Jans, R., & Degraeve, Z. (2007). Meta-heuristics for dynamic lot sizing: A review and comparison of solution approaches. *European Journal of Operational Research, 177*(3), 1855–1875.

Kämpf, M., & Köchel, P. (2006). Simulation-based sequencing and lot size optimisation for a production-and-inventory system with multiple items. *International Journal of Production Economics, 104*(1), 191–200.

Kang, J. H., & Kim, Y. D. (2010). Coordination of inventory and transportation managements in a two-level supply chain. *International Journal of Production Economics, 123*(1), 137–145.

Karimi, B., Ghomi, S. F., & Wilson, J. M. (2003). The capacitated lot sizing problem: A review of models and algorithms. *Omega, 31*(5), 365–378.

Kofler, M., Beham, A., Wagner, S., Affenzeller, M., & Achleitner, W. (2011). Re-warehousing vs. healing: Strategies for warehouse storage location assignment. *In 3rd IEEE International Symposium on Logistics and Industrial Informatics* (pp. 77–82). Piscataway: IEEE.

Lee, Y. H., Jung, J. W., & Lee, K. M. (2006). Vehicle routing scheduling for cross-docking in the supply chain. *Computers & Industrial Engineering, 51*(2), 247–256.

Li, M., Chen, X., & Liu, C. (2008). Pareto and niche genetic algorithm for storage location assignment optimization problem. *3rd International Conference in Innovative Computing Information and Control. ICICIC'08* (p 465). Piscataway: IEEE.

Liu, S. C., & Chen, J. R. (2011). A heuristic method for the inventory routing and pricing problem in a supply chain. *Expert Systems with Applications, 38*(3), 1447–1456.

Mateus, G. R., Ravetti, M. G., de Souza, M. C., & Valeriano, T. M. (2010). Capacitated lot sizing and sequence dependent setup scheduling: an iterative approach for integration. *Journal of Scheduling, 13*(3), 245–259.

Mishra, N., Kumar, V., Kumar, N., Kumar, M., & Tiwari, M. K. (2011). Addressing lot sizing and warehousing scheduling problem in manufacturing environment. *Expert Systems with Applications, 38*(9), 11751–11762.

Moin, N. H., & Salhi, S. (2007). Inventory routing problems: A logistical overview. *Journal of the Operational Research Society, 58*(9), 1185–1194.

Muller, L. F., Spoorendonk, S., & Pisinger, D. (2012). A hybrid adaptive large neighborhood search heuristic for lot-sizing with setup times. *European Journal of Operational Research, 218*(3), 614–623.

Nascimento, M. C., Resende, M. G., & Toledo, F. M. (2010). GRASP heuristic with path-relinking for the multi-plant capacitated lot sizing problem. *European Journal of Operational Research, 200*(3), 747–754.

Park, C., & Seo, J. (2010). Comparing heuristic algorithms of the planar storage location assignment problem. *Transportation Research Part E: Logistics and Transportation Review, 46*(1), 171–185.

Pitakaso, R., Almeder, C., Doerner, K. F., & Hartl, R. F. (2006). Combining population-based and exact methods for multi-level capacitated lot-sizing problems. *International Journal of Production Research, 44*(22), 4755–4771.

Robinson, P., Narayanan, A., & Sahin, F. (2009). Coordinated deterministic dynamic demand lot-sizing problem: A review of models and algorithms. *Omega, 37*(1), 3–15.

Silver, E. A., & Meal, H. C. (1973). A heuristic for selecting lot size quantities for the case of a deterministic time-varying demand rate and discrete opportunities for replenishment. *Production and Inventory Management, 14*(2), 64–74.

Sindhuchao, S., Romeijn, H. E., Akçali, E., & Boondiskulchok, R. (2005). An integrated inventory-routing system for multi-item joint replenishment with limited vehicle capacity. *Journal of Global Optimization, 32*(1), 93–118.

Speranza, M. G., & Ukovich, W. (1994). Minimizing transportation and inventory costs for several products on a single link. *Operations Research, 42*(5), 879–894.

Staggemeier, A., Clarke, A., Aickelin, U., & Smith, J. (2002). A hybrid genetic algorithm to solve a lot sizing and scheduling problem. *Internal Research Report MS-2002-4*. Bristol: University of the West of England.

Toptal, A., Koc, U., & Sabuncuoglu, I. (2013). A joint production and transportation planning problem with heterogeneous vehicles. *Journal of the Operational Research Society, 65*(2), 180–196.

Torabi, S. A., Ghomi, S. F., & Karimi, B. (2006). A hybrid genetic algorithm for the finite horizon economic lot and delivery scheduling in supply chains. *European Journal of Operational Research, 173*(1), 173–189.

van Hoek, R. I. (2001). The rediscovery of postponement. A literature review and directions for research. *Journal of Operations Management, 19*, 161–184.

Vidović, M., Popović, D., & Ratković, B. (2014). Mixed integer and heuristics model for the inventory routing problem in fuel delivery. *International Journal of Production Economics, 147*, 593–604.

Wagner, H. M., & Whitin, T. M. (1958). Dynamic version of the economic lot size model. *Management Science, 5*(1), 89–96.

Waters-Fuller, N. (1995). Just-in-time purchasing and supply: A review of the literature. *International Journal of Operations & Production Management, 15*(9), 220–236.

Woarawichai, C., Kullpattaranirun, T., & Rungreunganun, V. (2011). Inventory lot-sizing problem with supplier selection under storage space and budget constraints. *IJCSI International Journal of Computer Science Issues, 8*(2), 250–255.

Xie, J., & Dong, J. (2002). Heuristic genetic algorithms for general capacitated lot-sizing problems. *Computers & Mathematics with Applications, 44*(1), 263–276.

Zhao, Q. H., Chen, S., & Zang, C. X. (2008). Model and algorithm for inventory/routing decision in a three-echelon logistics system. *European Journal of Operational Research, 191*(3), 623–635.

Zhao, Q. H., Wang, S. Y., & Lai, K. K. (2007). A partition approach to the inventory/routing problem. *European Journal of Operational Research, 177*(2), 786–802.

Chapter 5
Scheduling

Abstract In this chapter we discuss scheduling problems and how methods from computational intelligence can be applied to them. We start with general considerations on scheduling problems and discuss variants and some simple solution concepts. After that some standard scheduling problems are discussed in more detail followed by a discussion of further scheduling problems relevant to logistics and supply chain management. After that we discus solution approaches from the field of computational intelligence with emphasis on encoding issues, especially in the context of using evolutionary algorithms. The paper ends with a discussion of the importance and success of using respective solution approaches especially from the area of metaheuristics.

5.1 Introduction

The production problems discussed in the previous chapter, deal with planning tasks on a rather rough level. The planning is done for a number of periods such as days, weeks, or months, and it is only planned what should be produced in what amount. The question of how the production is done, i.e. which particular resources are used, is mainly left out. Also the specific times of production activities such as hours and minutes are not considered.

When dealing with the planning on a fine time scale it is more important to make sure that the plans are consistent and feasible. For instance, the planning of the resources must respect their given availability and capacities over time. Also, the employees' working hours, absences, break times, and other required idle times must be considered in the same way as their capacities, e.g. being able to only process one item at a time. Working hours may also apply to machines e.g. because of maintenance breaks or defined servicing periods. Moreover, already planned or dispatched activities influence the availability of the respective resources over time.

The detailed planning of jobs, tasks or activities which require specific resources is denoted as scheduling. Instead of resources, the expression "machines" is frequently used in the scheduling terminology irrespective of whether genuine machines are

T. Hanne, R. Dornberger, *Computational Intelligence in Logistics and Supply Chain Management*, International Series in Operations Research & Management Science 244,
DOI 10.1007/978-3-319-40722-7_5

involved or not. The result of the planning process, the schedule, is a plan which includes the assignment of the jobs to resources and the specific times when the jobs are to be executed. In some cases it is sufficient to assign the jobs to a pool of resources (without the need to specify an individual resource) or to specify the sequence of jobs to be executed on a resource is adequate for their later execution. However, the basic idea is that jobs are planned on a detailed level which can be used for their final execution.

Scheduling problems arise in many different areas. Apart from the planning of production, logistics processes and other business processes schedules are also required in purely technical fields such as the automatic execution of jobs using IT infrastructure, e.g. planning the execution of jobs on a CPU. Other technical resources to be planned might be machines, service stations, equipment and tools, communication channels, or memory (see, e.g. Brucker and Brucker (2007) for a general introduction to scheduling problems and solution approaches).

Traditionally, scheduling problems in business were solved manually, i.e. a human assigned the jobs to resources and took care of finding a feasible time slot. Typically, tools such as large boards containing Gantt diagrams (or other forms of visualization) were used. The applied planning logic was rather a rule-of-thumb or an unspecified heuristics than a well-defined planning strategy or algorithm. Although the results might benefit from the experience of the planner, in no way optimal results were to be expected. This is hardly surprising taking into account that many scheduling problems are NP-hard (as we will discuss later in more details).

The advancement in computer technology and the necessity of automatic planning due to speed requirements (e.g. planning the execution of computer programs) fostered an increased usage of scheduling algorithms. Today, such algorithms are often used in areas which were dominated by purely human planning not so long ago.

5.2 Simple Rules and Heuristics

A scheduling problem requires specifying a number of characteristics of the jobs to be planned. For instance, job i can be described by the following data:

- arrival time at_i (time from when the job is available to be executed)
- processing time p_i (duration required for executing the job, may depend on the specific choice of a resource)
- due date dd_i (point in time by which the job should be completed)
- deadline dl_i (point in time after which accomplishing the job is not allowed or not possible)
- start time s_i (to be planned)
- end time (completion time) C_i (to be planned)
- possible priorities of the jobs

Moreover there can be dependencies between jobs, especially dependencies concerning feasible sequences of jobs. The most common situation is that one job must be completed before another job can be started (end-start-relationships). For instance, in painting the base coat must be finished before the final coating is done. Although end-start-relationships might be the most common relationships between jobs, there may also be start-start-relationships (one job can start only after another has started), start-end-relationships (one job must have started before another can be completed), or end-end-relationships (a job can be finished only after another job has been completed).

All these relationships can require additional minimum or maximum times to be considered. For instance, one job must be finished at least x hours before another job can start. In the example mentioned above, the base coat must have dried (which requires a particular time) before the final coating can be applied.

This example considers a minimum time required between two jobs. An example of a maximum time could be the following: When insulation boards are affixed, this should happen before the adhesive has dried. In this case, a job must start not later than x hours after another job has been completed.

Further planning restrictions may result from the required resources. Especially, some resources are suitable only for a particular type of job. For instance, an employee needs a particular qualification for performing a given task, e.g. driving a vehicle.

Moreover, some jobs may require several resources successively or in parallel, e.g. a machine and an operator. Resources can have different capacities, e.g. machines which can service several items at the same time.

Finally, the temporal availability of resources must be considered (available capacities over time e.g. because of shifts, breaks, maintenance times, already dispatched jobs, etc.).

For scheduling problems various objectives could be relevant: Typical objectives may refer to small throughput times, a fair balancing of loads among the resources, a high throughput, or the adherence to delivery dates.

For each of these general aspects there are usually different ways how the objectives can be specified. For instance, the throughput time of a single job can be defined as the duration from the arrival of the job (arrival time) until it is finished (finishing time). For a set of jobs to be planned, one could consider then the average throughput time or the maximum throughput time of the considered jobs. This maximum throughput time is also called length of a schedule or simply makespan C_{max}. Also the throughput time of all jobs together (sum of throughput times) could be considered.

With respect to the adherence to delivery dates, for a single job, one could consider its tardiness (lateness) and/or earliness. A job is late when it is completed after its due date ($c_i > dd_i$). It is early, when the due date is after its finishing time ($c_i < dd_i$).

For a set of jobs, the maximum tardiness

$$max\{c_1 - dd_1, \ \ldots, \ c_n - dd_n, \ 0\} \tag{5.1}$$

and/or maximum earliness

$$max\{dd_1 - c_1,\ \ldots,\ dd_n - c_n,\ 0\} \tag{5.2}$$

could be of interest or the sum of tardiness L of delayed jobs

$$la = \sum_{i=1}^{n} \max\{c_i - dd_i, 0\} \tag{5.3}$$

and/or the sum of earliness E of early jobs

$$ea = \sum_{i=1}^{n} \max\{dd_i - c_i, 0\}. \tag{5.4}$$

If earliness and tardiness are both important, one could possibly consider both and sum up the respective values as an unpunctuality indicator P possibly with weighing tardiness stronger if it is more important to avoid lateness than earliness:

$$P = w_L \cdot la\ +\ w_E \cdot ea,\ with\ 0 < w_E \leq w_L. \tag{5.5}$$

Another approach might be to calculate simply the number of late or early jobs.

With respect to a good utilization of resources, we could try to process as many jobs as possible during a given time interval (maximizing performance) or we could try to plan a given set of jobs with as few resources as possible (minimizing costs).

Load balancing might be motivated by fairness considerations, esp. when the resources are employees. But also due to technical reasons, it might be useful to utilize given machines in an approximately even way.

Further general aspects of scheduling problems are how they are solved with respect to the available information and what basic requirements for solutions are to be considered.

With respect to the latter, a rough distinguishing characteristic of scheduling problems is whether it is allowed to adjourn jobs. If the interruption of jobs is allowed the planning is also called preemptive scheduling. A typical situation for preemptive scheduling is the processing of jobs on a computer. Typical multi-tasking operating systems assign tiny time slices to each job after which the jobs are put into a queue, waiting for their next time slice. That means, the jobs are actually processed sequentially (on a single processor or single core system) and only due to short intermediate waiting times the impression of parallel execution is created.

In the case of non-preemptive scheduling it is not allowed to interrupt an already started job. In some areas, there may be some technical or organizational necessity for that. In other situations, adjourning jobs may be possible, but as stopping and re-starting a job may include significant setup times (and costs) this may be rather undesirable. For instance, interrupting one transport, doing another transport, then continuing with the first transport requires additional driving distances and time.

Interrupting a production process may require cleaning operations, a change of tools, or simply the adaption of a worker to new tasks. When generating a schedule, therefore, interruptions should be avoided.

Another basic question is whether all required information is available by the time of scheduling. Usually, for simple scheduling problems it is assumed, that all jobs and other required information are known when the schedule is generated. In practice, however, such information is often known only partially and may be subject to changes over time. For instance, only a part of the jobs may be known when the schedule for the next day is prepared. New jobs (e.g. due to new customer orders) may then arise over the day. Possibly, some jobs are cancelled or postponed. Or the parameters of the jobs are not known or are subject to changes. There may occur, for example, unexpected delays in the processing times.

All these practical aspects turn the (static) scheduling problem (also called offline scheduling problem) into a dynamic one, also known as online scheduling problem (Ouelhadj and Petrovic 2009; Pruhs et al. 2004; Albers 2003). Coping with a dynamic scheduling situation may follow different strategies. One extreme approach would be to consider the given information and treat the problem like a static one. If new information arises (e.g. new jobs to be planned) a completely new schedule is calculated (without considering the previous plan). Another extreme approach would be to adhere to all planned activities and only somehow add the new job to the current plan (or jobs with changed information are rescheduled, e.g. because of delays of previously scheduled jobs). Both approaches have their different advantages and disadvantages. The first approach allows for a completely new plan, optimized according to the new information. It is, however, more costly in terms of planning effort and computation times and may lead to frequent and drastic changes from plan to plan. The second approach is less costly and leads to schedules which are more robust over time. However, the results may be poor with respect to the planning objectives. A better solution may be based on a "compromise" between those two ideas. For instance, only particular parts of the plan remain as planned (e.g. "freezing" the jobs scheduled for the near future) whereas the others are rescheduled. Or new plans are derived in the "neighborhood" of the current schedule (e.g. by a neighborhood search approach).

5.3 Standard Scheduling Problems

As discussed above, scheduling problems can be distinguished by various characteristics. Based on a proposal by Graham et al. (1979) the following classification approach has found wide acceptance for describing different types of scheduling situations: A scheduling problem is denoted by a 5-tuple $\alpha/\beta/\gamma/\delta/\varepsilon$ where each of the Greek letters is a placeholder for one aspect of the scheduling problem:

- α denotes the number of jobs
- β denotes the number of resources (machines) with limited capacity

- γ refers to the constraints (e.g. arrival times of jobs, due dates, dependencies between jobs)
- δ refers to the processing times of jobs (which may depend on the specific assignments of resources)
- ε refers to the objective function (or objective functions if several are to be optimized)

For some types of scheduling problems such as single machine scheduling problems ($\beta = 1$) simple rules may be available to optimize the considered objective functions. For instance, if the average throughput time (or flow time) is to be minimized, the so-called Shortest Processing Time (SPT) rule (also denoted as shortest job first strategy) leads to optimal results. The rule requires that jobs (ready to be processed) are carried out in the sequence of non-decreasing processing times, i.e. the job with the shortest processing time is executed first.

Example: Assume there are three jobs to be carried out with processing time 60 min, 20 min and 10 min. The execution of the jobs in that sequence leads to an average throughput time of (60 min + 80 min + 90 min)/3 = 76.66 min. Carrying out the jobs in the inverse order, i.e. according to the SPT rule, leads to an average throughput time of (10 min + 30 min + 90 min)/3 = 43.33 min.

A disadvantage of this rule is that long jobs (having a long processing time) may "starve". This may happen when continuously new jobs arrive during the processing of already available jobs. The rule would require that newly arriving jobs should be executed before already waiting jobs if they have shorter processing times. Therefore, it may take very long until the jobs with long processing times are finally completed.

Obviously, the SPT does not minimize the maximum throughput time of jobs. This objective is optimized when the jobs are executed as early as possible in the sequence of their arrival times. This rule may also take care of dependencies between jobs. In this case, jobs are to be carried out as early as possible in the sequence of their arrival times when their predecessor jobs are planned.

A similar rule can be applied for minimizing the maximum lateness of a set of jobs (Earliest Due Date-Scheduling) (Jackson 1955). Assuming that there are no dependencies between jobs and that all jobs are available at the same time (or that they can be interrupted), the jobs are to be executed in the sequence of non-decreasing due dates.

In case of dependencies between jobs a modification of that rule can be applied, which makes sure that jobs are only planned when their predecessors have been executed.

The most often considered harder scheduling problems are the job shop scheduling problem, the flow shop scheduling problem, and the open shop scheduling problem which are discussed in the next subsections.

5.3.1 *Job Shop Scheduling*

The job shop scheduling problem can be characterized as follows: There are *n* jobs to be scheduled on *m* different machines. Each job consists of a set of *m* operations (corresponding to each of the machines) and the sequence of these operations is pre-specified but not necessarily identical for all jobs. Each operation is characterized by the required machine and a given processing time.

There are further constraints on the jobs and machines, such as,

- a job does not visit the same machine twice;
- there are no dependencies between operations of different jobs;
- operations may not be interrupted (non-preemptive scheduling);
- each machine can process only one job at a time;
- all jobs are available by the same time;
- due dates are not specified.

Figure 5.1 shows a small example of a job shop scheduling problem with three jobs to be processed on three machines. In the example, the machines are to be used in given sequences which are different for each job. For instance, job 3 needs to be processed on machine 2 first, then on machine 1, and finally on machine 3. The problem specification also includes the specification of the processing times of each operation.

The Gantt diagrams in Fig. 5.2 show two possible solutions for the example problem. Note that the second solution leads to a smaller makespan, i.e. a shorter completion time for all three jobs.

The usual objective is to minimize the makespan, i.e. the time required to complete all jobs. It was proved that this problem is NP-hard (Garey et al. 1976).

	Sequence of machines and processing times		
Job 1	1, 4 min	2, 3 min	3, 2 min
Job 2	1, 4 min	3, 4 min	2, 3 min
Job 3	2, 4 min	1, 3 min	3, 2 min

Fig 5.1 Example of a job shop scheduling problem with three jobs consisting of three operations

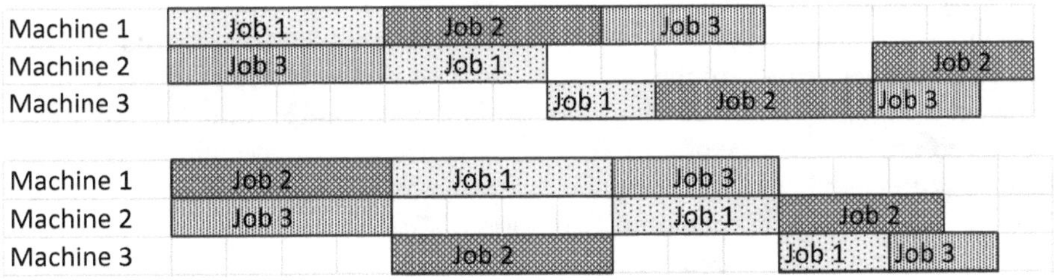

Fig 5.2 Two Gantt diagrams of possible solutions to the problem from Fig. 5.1

An important generalization of the above job shop scheduling problem is given by the flexible job shop scheduling problem which allows an operation to be processed on more than one machine, i.e. the machines may be capable of performing more than one type of operation (see, e.g., Gambardella and Mastrolilli 2000). Like in the job shop scheduling problem there is at least one machine available for the processing of each operation. Due to choices concerning the machines used for an operation, one could also call the flexible job shop scheduling problem an assignment problem (determination of the machines used for each operation) in combination with a traditional job shop scheduling problem. This problem is also NP-hard.

Considering modern production technology and the idea of flexible manufacturing systems (Basnet and Mize 1994), the flexible job shop scheduling scenario is usually considered more realistic than the traditional job shop scheduling problem. Another realistic variant of the problem is the flexible job shop scheduling problem with work centers which assumes that resources for identical activities (e.g. pressing or welding) are organized as a group (i.e. work centers) and where the first and the last work center to be used for a job is pre-specified (Behnke and Geiger 2012).

5.3.2 *Flow Shop Scheduling*

The flow shop scheduling problem is a special case of the job shop scheduling problem characterized by a given sequence of operations which applies to all jobs. There are n jobs to be scheduled consecutively on m different machines (sometimes also called stages). As for the job shop scheduling problem, processing times for each job on each machine are given, jobs are not to be interrupted, machines can handle only one job at a time, etc.

Expressed mathematically, each of the n jobs from the set $J = \{1, \ldots, n\}$ has to be processed on m machines in the order $\{1, \ldots, m\}$. Each job $j, j \in J$, consists of a sequence of m operations, each of them corresponding to the processing of job j on machine i during an uninterrupted processing time $p_{ij} > 0$. The objective of the flow shop scheduling problem is usually the minimization of the makespan, although other objectives or a multiobjective version might certainly be practically relevant.

Under suitable conditions, it can be shown that it is sufficient to consider only permutation schedules. A permutation schedule is one in which jobs are processed in the same order on each of the machines and a machine is not kept idle if a job for the machine is ready for processing. The solution of the problem can then be simply expressed as a permutation of the jobs denoted as $\pi = (j_1, \ldots, j_n)$ with $j_i \in \{1, \ldots, n\}$ indicating the number of the job to be processed at ith position on each machine. The completion time of operation k from job j_i, $C(j_i, k)$, can then be calculated as follows (see, e.g., Liu et al. 2007):

$$
\begin{aligned}
C(j_1, 1) &= p_{1j_1}, \\
C(j_i, 1) &= C(j_{i-1}, 1) + p_{1j_i}, \\
C(j_1, k) &= C(j_1, k-1) + p_{kj_1}, \\
C(j_i, k) &= \max\{C(j_{i-1}, k), C(j_i, k-1)\} + p_{kj_i},
\end{aligned}
$$

for $i=2, \ldots, n$, and $k=2, \ldots, m$. The resulting makespan is

$$C_{max}(\pi) \ = C(j_n, m).$$

Note that there are cases of the general flow shop scheduling problem where permutation solutions are not optimal (Dudek et al. 1992). In solution approaches, however, often only solutions described by permutations of jobs are considered. Therefore, a respective version of the flow shop scheduling problem restricting allowed solutions to permutations is also denoted as permutation flow shop scheduling problem.

Another variant (and generalization) of the flow shop scheduling problem is the hybrid flow shop scheduling problem. In this problem there are several stages of the jobs (corresponding to consecutive operations) where several machines may be available for each stage. Thus, this problem allows for some more flexibility concerning the processing of operations similar to the flexible job shop scheduling problem. At the same time, the problem can be considered a generalization of the parallel machine scheduling problem (Ribas et al. 2010). Hybrid flow shop problems exist in various subvariants, e.g. with identical machines at each stage, with uniform machines characterized by different speeds, or by so-called unrelated machines which are characterized by the fact that the processing times are specific to each machine at a given stage.

Another often studied variant of the flow shop problem is called continuous flow shop scheduling problem or no-wait flow-shop scheduling problem (Czogalla and Fink 2012) and requires that jobs be processed without intermediate waiting time, i.e. when a job is started it must be made sure that when finishing one machine, the next machine is already available.

5.3.3 Open Shop Scheduling

The open shop scheduling problem is similar to the job shop and the flow shop situation: It is assumed that there are n jobs to be scheduled on m different machines. However, the open shop scheduling problem is characterized by the lack of any constraints on the sequence of operations. Each job needs to be processed by each machine and the respective processing times are given. As before, the jobs are to be scheduled in a non-preemptive way and each machine can handle only one job at any time (and vice versa each job can be treated only by one machine at a time).

In most studies of the problem the objective is the minimization of makespan—just as for the above problem types. In some papers, the sum of completion times of jobs (or, equivalently, the average completion time) is considered a minimization objective. Both problem variants are NP-hard in the general case for $m > 2$ (Andresen et al. 2008)—just like the job shop and the flow shop problems. However, due to much more freedom with respect to possible operation sequences on the machines, the search space of the open shop problem is much larger which induces difficulties especially for solution approaches that rely on a smart exploration of the search space, e.g. branch & bound approaches. In contrast, algorithms which make use of parallelism might benefit from that situation (Prins 2000).

From a practical point of view, open shop problems may be considered more applicable than the previous two standard problems, the job shop and the flow shop problem in their canonical versions. For instance, work-center based manufacturing, maintenance, or repair often allow a significant amount of freedom concerning the sequence of operations. Though, some restrictions on such sequences may still apply and, moreover, multiple resources suitable for particular operations may be available in a practical setting. Thus, a variant of an open shop scheduling problem or one of the above variations of standard scheduling problems may be more appropriate for modelling such practical situations.

5.4 Specific Scheduling Problems in Logistics

Let us note that there are many other types of scheduling problems which may be relevant for logistics applications. Relevant examples can certainly be found in the field of resource constrained project scheduling (Herroelen et al. 1998; Brucker et al. 1999). Such problems reflect general scheduling situations with scarce resources and jobs with different kinds of dependencies as they may appear in production, construction, project work but also for specific logistics activities. Other standard scheduling problems such as the job shop scheduling problem may be considered as special types of the resource constrained project scheduling. The richness of resource constrained project scheduling problems may also consider, for instance, different modes in the sense of processing alternatives for jobs or different types of resources such as renewable and nonrenewable resources.

Another scheduling area with significant relevance for logistics practice is definitely the scheduling of staff which is also called rostering. The problem assumes that there are various possible time slots where employees could work and that a respective planning is necessary (in contrast to self-scheduling or self-management which does not assume a central planning logic). Rostering problems occur mostly where employees work shift, e.g. in hospitals, in large industrial plants, or at transport service providers such as train companies. In some areas such problems can be formulated with clearly defined time slots for the shifts. In other areas, e.g. rail operations, the time slots may vary over time since services cannot be interrupted arbitrarily. Instead, there might be specifications of maximum shift durations and rules for compensating long shifts with shorter shifts or days off.

Apart from such complications resulting from the requirement of the underlying work activities, rostering problems may include further constraints, e.g. concerning the feasibility of particular shift assignments, the total number of assignments of a person, the time between two subsequent assignments, or the handling of particular assignments (e.g. night shifts). The objective function could include a measure for the violation of general constraints or consider individual preferences of the employees.

A similar type of scheduling problems with defined time slots for activities is found in the area of time-tabling. Here, we usually have fixed slots for activities of equal or different length, and the activities have to be assigned to suitable time slots, often for a recurring purpose. Typical examples are the time-tabling problems in schools, universities, or other educational institutions. Activities like courses have to be carefully planned to avoid time conflicts (like overlapping assignments) for participants such as students, classes, and teachers. Also other resources such as rooms or other infrastructure have to be considered, and further requirements may apply, e.g. the minimum or maximum temporal distances between planned activities or the maximum or minimum time assigned to a person per day.

In specific operations like in transport management, scheduling and rostering tasks are linked up with other activities such as transport operations. This means that scheduling does not only consider the allocation of resources over time but also needs to consider the duration of transport and, as a consequence, the change of locations. For instance, transport times result from the distances between locations, the assumed speed, and the time required for related purposes, e.g. for breaks or transshipment activities. In the overall planning of activities it might be necessary to bring resources (vehicles, drivers, etc.) first to the places of activities (e.g. pickup locations) and to return them to the defined locations (e.g. depots) after having finished the duties. Due to these special situations, transport planning activities are usually considered quite different scheduling problems and are usually referred to as routing or vehicle routing (see Chap. 3).

Different scheduling problems arise for line operations providers. These transport services have a recurring structure which is based on arriving at different locations in a sequence also called a line. In line transport, there are often timetables which serve these locations with specific cyclical structures, e.g. one train per hour. The planning of line operations and the elaboration of timetables requires considering, among other things, the expected transport requirements, the available resources and costs, and the required time for locomotion, stops, and other purposes.

A further complication in the planning results from the coordination of different lines: A larger network of locations can be served by several lines which allow passengers or goods to change from one line to another. It is advisable to avoid unnecessary waiting times at transfer locations, e.g. by having two means of transportation that stop at the same time at a common location and wait until the transfer is finished. This concept, however, has some disadvantages besides the required transfer time: If one means of transport is late, the operations of the other means are late too if it waits for the respective connection. This may cause further lateness for subsequent connections and, possibly, lateness is finally spread over the whole network. If waits are not stipulated in the operations, lateness may often lead to missed connections. Thus, delay management can be a major issue in the planning of such transport networks and the calculation of timetables (see, e.g., Liebchen et al. 2010).

5.5 Solving Scheduling Problems with Computational Intelligence Techniques

The discussed standard scheduling problems (job shop, flow shop, and open shop) can only be solved efficiently in special cases which usually correspond to the problem size (Garey et al. 1976; Williamson et al. 1997). For instance, the open shop scheduling problem can be solved by a polynomial algorithm if there are either only two jobs or two machines or if all processing times are identical. In general, the problem is NP-hard which makes it a good candidate for applying metaheuristic algorithms.

As a consequence, there is a large amount of publications dealing with metaheuristics and other computational intelligence techniques for solving scheduling problems. While fuzzy approaches are often used to model some vagueness in scheduling problem formulations (e.g. imprecise processing times, fuzzy due dates, or flexible constraints, see, e.g., Dubois et al. (2003)) neural networks seem to play a minor role in scheduling. In the following subsections, we focus on a suitable representation of scheduling problems as they are needed for applying solution techniques and add some information on the current importance of employing metaheuristics.

5.5.1 Encoding Issues

For treating scheduling problems with a suitable optimization routine or heuristic it is usually not required to determine the exact times for the execution of jobs or operations. Instead, it is mostly sufficient to determine the sequence of execution (when being assigned to a particular machine). With this information it is not difficult to calculate the exact times assuming that the jobs are executed in the given sequence as early as possible. For instance, the first job is executed as early as possible, i.e. not before its arrival time and as soon as the machine is available (there might be other jobs already fixed for that machine). Knowing the processing time of the job one can calculate when the machine becomes idle again (also here possibly taking into account already fixed jobs). For that time, the second job in the sequence can be started—or later when it is not available by that time and so on. (See above for an example of that approach in the case of flow shop scheduling in Sect. 5.3.2.)

Such a procedure for calculating start and end times of jobs could also consider precedence relationships if not already incorporated in the sequence. For instance, in a job shop scheduling problem, one can account for the second operation of a job that it cannot be started before the first operation is completed. This can be easily achieved when the concrete schedules for the different machines are set up step by step always continuing with the next machine that becomes idle (In the beginning when all machines may be idle the sequence of considered machines does not matter).

Representing schedules in this more abstract form (i.e. only encoding assignments to machines and the sequence of executed jobs per machine) has significant advantages. The search space is much smaller and the problem of calculating infeasible solutions by modification operations (e.g. mutations in an evolutionary algorithm) is relieved. There is, of course, the disadvantage that evaluating a solution (e.g. calculating the makespan) takes additional time.

Example Job 1: operation 1: 8 min; operation 2; 10 min
Job 2: operation 1: 15 min; operation 2: 5 min
Job 3: operation 1: 5 min; operation 2; 10 min
machine 1; job 1; job 2; job 3
machine 2: job 3; job 1; job 2
This leads to the following schedule:
machine 1: Job 1: 0–8 min, Job 2: 8–23 min; Job 3:23–28 min.
machine 2: Job 3: 0–10 min; Job 1: 10–20 min; Job 2: 23–28 min

For the above reason, the representation of a scheduling problem solution requires mainly two things: the assignment of specific machines for tasks or operations (if not clear beforehand as e.g. in job-shop scheduling problems) and the sequence of tasks or operations at the assigned machines. For instance, there is an integer array of length o with values $j \in \{1, \ldots, m\}$ at position i saying that operation $i \in \{1, \ldots, o\}$ is assigned to machine j. Another possibility is that the values j are index values representing the jth machine from those being feasible for operation i (Fig. 5.3) (cf. Zhang et al. 2010).

The sequence-specific part of a solution representing sequences or permutations of objects (tasks, operations) could be constructed as follows in a straightforward way: For instance, we could represent the objects 1, ..., 5 by a string of size 5, e.g. (2, 4, 5, 3, 1) when task 2 is executed first, then task 4, and so on. A respective representation could also possibly be simplified: For instance, in Zhang et al. (2010) for the case of a flexible job shop scheduling problem only the job number is indicated for the sequence of operations but not the individual operations, because the sequence of operations within a job is predefined.

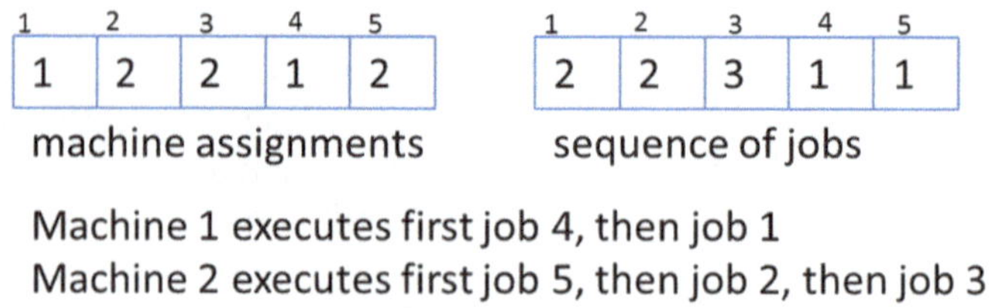

Fig 5.3 Example of the encoding of a scheduling problem solution with five jobs. The first array defines the machine assignments of the jobs whereas the second array defines the sequence for the respective machines

Another possibility to represent a schedule, e.g. for open shop problems, is the rank matrix with its advantages of avoiding redundancy and providing a unique representation of solutions (Andresen et al. 2008). An approach based purely on sequences is investigated in Prins (2000) for the case of the open shop scheduling problem. The sequence of jobs can be interpreted as a priority list. Based on this priority list and using specific simple heuristics a schedule is built. For instance, a non-delay heuristic is used for planning the next job or a heuristic based on the first-in first-out (FIFO) principle is applied.

However, such a direct representation of sequences may be disadvantageous within a search strategy: Let us consider a small change of a solution, e.g. changing the last item in the sequence from 1 to 2 due to a random mutation. The resulting representation (2, 4, 5, 3, 2) corresponds to an infeasible solution as "2" occurs twice in the sequence whereas "1" does not occur at all. At least two "synchronized" changes in the sequence are therefore necessary to obtain again a feasible solution.

An alternative representation of a sequence denoted as random key encoding helps to avoid such feasibility problems (see Bean 1994): Let us assume that the sequence of items is represented by real values, e.g. (1:0.75, 2:0.08, 3:0.72, 4:0.33, 5:0.51). The real value at position i in the vector denotes its "priority" in the sequence. Sorting the items according to this number leads to the represented sequence (2:0.08, 4:0.33, 5:0.51, 3:0.72, 1:0.75). In this example we assume that lower values indicate an earlier execution of a task.

A change of these values always leads to another feasible solution and small changes of the value usually lead to small changes in the solution, e.g. reducing the values for item 1 by 0.1 swaps it with item 3 while an increase does not have any effect at all.

Also when combinations of solutions are calculated (like by recombination or crossovers in an evolutionary algorithm), the result usually becomes infeasible when the direct sequence representation is used, whereas random key encoding helps to preserve feasibility.

Random key encoding is often used in the framework of genetic algorithms (Bean 1994; Mendes et al. 2009; Gonçalves et al. 2011) but can also be used conveniently in connection with other metaheuristics, e.g. differential evolution (Nearchou and Omirou 2006), local search (Gonçalves et al. 2005), variable neighborhood search (Roshanaei et al. 2009), particle swarm optimization (Lin et al. 2010; Kuo et al. 2009) or memetic algorithms (Liu et al. 2007).

While random key encoding may preserve feasibility with respect to maintaining a well-defined sequence during variation operations in the search algorithms, further constraints such as precedence constraints are not automatically observed. Therefore, one might think of more advanced operators which take care of feasibility in a more general sense.

Let us first consider again the simpler problem of representing sequences and keeping them feasible during the crossover or recombination in an evolutionary algorithm. We assume that the sequences are represented "directly", i.e. for representing the sequence of n jobs, we have a chromosome of length n with entries representing the job indexes, i.e. the first entry is the index of the job executed first and so on. In the example in Fig. 5.4 two parent solutions representing sequences of eight jobs are to be recombined by a one-point crossover. After determining a random crossover position (after chromosome 4 in the example), the first part of each parent solution is copied into the respective part of an offspring solution. Unlike in regular crossover, the second part of the offspring solutions is not filled up with data from the second parent at the same chromosome locations. Instead, the second parent solution is checked (starting at the crossover location and going to the right) for entries which are not yet included in the first part of the offspring to fill up the remaining chromosome locations. When the end of the chromosome is reached the process is continued at the first position. Finally, each number {1, …, 8} appears exactly once in each offspring solution.

In the example in Fig. 5.5 two parent solutions representing sequences of eight jobs are to be recombined by a two-point crossover OX. First, two crossover positions are determined randomly. The middle part between the two crossover locations is copied from each parent solution to a corresponding offspring solution. The remaining parts are filled up similar to the case of the one-point crossover in

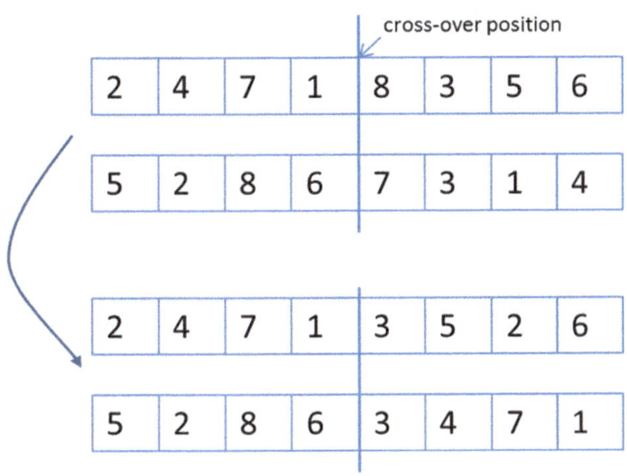

Fig 5.4 Example of a one-point crossover for a sequence of eight jobs

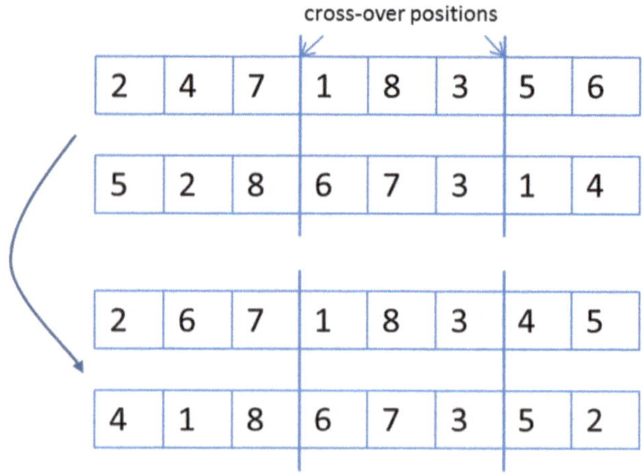

Fig 5.5 Example of a two-point crossover OX for a sequence of eight jobs

Fig. 5.4. That means, starting at the second crossover point (and going to the right), it is checked whether the entries are already included in the offspring solution. If not, they are used for the current position which has to be filled in the offspring solution. When the end of the parent chromosome is reached the process continues with the first position. Similarly, when the last offspring entry is filled, the process continues with the first position to fill also the part before the first crossover location.

Let us note that these operators work quite efficiently. From an algorithmic point of view the construction of new solutions and the control, in particular, whether no jobs have been forgotten or been represented twice in a sequence are done simultaneously without relevant extra effort. An array can be used to keep track for each job index whether it is already included in an offspring solution or not. Therefore, the construction of a new solution does not require repeated searches for checking whether a currently considered entry has already been included.

When we consider more complex scheduling problems with further constraints such as precedence constraints, similar concepts for feasibility preservation can be utilized. In particular, two encoding approaches and related operators are used in the framework of evolutionary algorithms: the precedence preserving order-based crossover (POX) and precedence preserving shift mutation (PPS) (Lee et al. 1998; also cf. Pezzella et al. 2008). The POX operation works with two solution representations (parents) and derives as is usual in crossover two offspring solutions. First, an operation from the first parent is selected and copied to the first offspring solution together with all the other operations belonging to the same job. After that, the solution is completed with the remaining operations, in the same order as represented in the second parent. For the second offspring solution the process is applied analogously. By doing so POX preserves the precedence constraints.

Let us note that there are several other recombination operators in use which take care of the feasibility of resulting solutions, e.g., the partial-mapped crossover

(PMX) or the order crossover (OX) discussed above which can be interpreted as two-point crossovers (exchange of randomly selected substrings) with a subsequent repair mechanism (Cheng et al. 1999). Further recombination operators used occasionally in the literature are position-based crossover, order-based crossover, cycle crossover (CX), linear order crossover (LOX), subsequence exchange crossover, job-based order crossover, partial schedule exchange crossover, substring exchange crossover (Cheng et al. 1999). Let us mention that crossover operations for sequence-based representations for scheduling problem solutions are similar to respective operations in the context of the TSP, cf. Sect. 3.4.2.

PPS (Lee et al. 1998) selects an operation from a single parent chromosome and shifts it to another position while taking care of the precedence constraints for that operation. Other possibilities for (random) changes are inversion mutation where a substring defined by two random locations is inverted in the chromosome. Similarly a random substring could be shifted to a new random position. A further mutation approach is the exchange mutation where the genes from two random positions are exchanged (Cheng et al. 1999).

Another important prerequisite for applying various metaheuristics is the definition of a neighborhood, e.g. for local search, tabu search, or variable neighborhood search (Gambardella and Mastrolilli 2000; Zhang et al. 2007, 2008, 2010). A small neighborhood can be defined by moving one operation. Since solutions with shorter makespan (throughput time) are sought one can restrict the neighborhood to movements of operations on the critical path. Larger neighborhoods (e.g. as required for variable neighborhood search), could be defined by the movement of two operations from a critical path, e.g. in the form of two swap operations (Zhang et al. 2007; Gao et al. 2008).

Another neighborhood search approach involving two neighborhoods (an iterative descent method) is as follows: One neighborhood is defined by reassigning a job from a machine with maximum makespan to another machine. The other neighborhood is defined by swapping the machine assignment of two jobs. In the iterative descent metaheuristic first successive improvements are sought in the first neighborhood type. When no further improvements occur, successive improvements by exploring type 2 neighborhoods are searched for.

5.5.2 *Usage of Metaheuristics in Scheduling*

Basically, job shop scheduling problems have been treated with almost all of the better known CI techniques and even some of the more obscure ones. Due to these extensive existing studies of scheduling problems we do not try to cover that aspect in details.

A short look at the literature reveals that among the most dominating metaheuristics we find genetic algorithms which are used, for instance, in Della

Croce et al. (1995). First applications of genetic algorithms date back to the 1980s and thus belong also to the oldest metaheuristics used for solving scheduling problems (see, e.g., Davis 1985).

Ruiz and Vázquez-Rodríguez (2010) present an interesting overview of the utilization of different solution techniques for the hybrid flow shop scheduling problem. According to their investigations, specific heuristics and dispatching rules are dominating in publications dealing with respective problems (50 %), while mathematical programming and branch & bound approaches are at the second position (25 %). Among metaheuristics applications, genetic algorithms are most commonly used (8 %). At positions 2 and 3 among the metaheuristics are tabu search (6 %) and simulated annealing (5 %). It should be noted that the study covers publications dating back to 1970 and that the utilization of metaheuristics has significantly increased during the last 15 years.

In an overview of recent approaches to the flexible job shop scheduling problem (Ye and Ma 2015) starting in 2005 ten of 25 approaches (40 %) are based on evolutionary algorithms (especially genetic algorithms), three are based on ant colony optimization (12 %), another three are based on other metaheuristics (12 %), while nine are based on mathematical approaches or problem-specific heuristics (36 %). Thus, in the newer survey, 64 % of the approaches are based on metaheuristics.

Unfortunately, it is not easy to find similarly compiled results for other types of scheduling problems indicating the usage, frequency and distribution of different classes of methods. However, the results for the flexible job shop scheduling problem do not appear to be untypical for the current importance of metaheuristics, whereas older bibliographies (e.g. Hoogeveen et al. 1997) show a clearly smaller share of metaheuristics.

Solution algorithms to scheduling problems are often tested using established benchmark instances such as those provided by Taillard (1993). Although this is still one of the most important sources of scheduling problems for comparing different algorithms, various novel collections have been added by many different authors. Although often identical test problems are used in studies of different metaheuristics (and other types of methods) it is not easy to derive general results which methods might be most suitable (especially with respect to performance). The usually different test settings are one reason (e.g. computer hardware). Another reason is that the performance largely depends on the adaptation of the method to the problem particularities like encoding, adaptation of operators, and parameter settings. There seems, however, to be some evidence that for some types of problems neighborhood search oriented approaches (like tabu search) seem to work best, whereas for others parallel searching approaches like genetic algorithms seem to have advantages. Anyhow, there are clear indications of the efficacy of some metaheuristics, because they allowed to solve test problems which have been practically unsolvable before or to decrease the solution times.

References

Albers, S. (2003). Online algorithms: A survey. *Mathematical Programming, 97*(1–2), 3–26.

Andresen, M., Bräsel, H., Mörig, M., Tusch, J., Werner, F., & Willenius, P. (2008). Simulated annealing and genetic algorithms for minimizing mean flow time in an open shop. *Mathematical and Computer Modelling, 48*(7), 1279–1293.

Basnet, C., & Mize, J. H. (1994). Scheduling and control of flexible manufacturing systems: A critical review. *International Journal of Computer Integrated Manufacturing, 7*(6), 340–355.

Bean, J. C. (1994). Genetic algorithms and random keys for sequencing and optimization. *ORSA Journal on Computing, 6*(2), 154–160.

Behnke, D., & Geiger, M. J. (2012). *Test instances for the flexible job shop scheduling problem with work centers.* Working Paper. Hamburg: Helmut-Schmidt-Universität.

Brucker, P., & Brucker, P. (2007). *Scheduling algorithms* (5th ed.). Berlin/Heidelberg: Springer.

Brucker, P., Drexl, A., Möhring, R., Neumann, K., & Pesch, E. (1999). Resource-constrained project scheduling: Notation, classification, models, and methods. *European Journal of Operational Research, 112*(1), 3–41.

Cheng, R., Gen, M., & Tsujimura, Y. (1999). A tutorial survey of job-shop scheduling problems using genetic algorithms, part II: Hybrid genetic search strategies. *Computers & Industrial Engineering, 36*(2), 343–364.

Czogalla, J., & Fink, A. (2012). Fitness landscape analysis for the no-wait flow-shop scheduling problem. *Journal of Heuristics, 18*(1), 25–51.

Davis, L. (1985). Job shop scheduling with genetic algorithms. In *Proceedings of an International Conference on Genetic Algorithms and their Applications* (Vol. 140). Pittsburgh, PA: Carnegie-Mellon University.

Della Croce, F., Tadei, R., & Volta, G. (1995). A genetic algorithm for the job shop problem. *Computers & Operations Research, 22*(1), 15–24.

Dubois, D., Fargier, H., & Fortemps, P. (2003). Fuzzy scheduling: Modelling flexible constraints vs. coping with incomplete knowledge. *European Journal of Operational Research, 147*(2), 231–252.

Dudek, R. A., Panwalkar, S. S., & Smith, M. L. (1992). The lessons of flowshop scheduling research. *Operations Research, 40*(1), 7–13.

Gambardella, L. M., & Mastrolilli, M. (2000). Effective neighborhood functions for the flexible job shop problem. *Journal of Scheduling, 3*(1), 3–20.

Gao, J., Sun, L., & Gen, M. (2008). A hybrid genetic and variable neighborhood descent algorithm for flexible job shop scheduling problems. *Computers & Operations Research, 35*(9), 2892–2907.

Garey, M. R., Johnson, D. S., & Sethi, R. (1976). The complexity of flowshop and jobshop scheduling. *Mathematics of Operations Research, 1*(2), 117–129.

Gonçalves, J. F., de Magalhães Mendes, J. J., & Resende, M. G. (2005). A hybrid genetic algorithm for the job shop scheduling problem. *European Journal of Operational Research, 167*(1), 77–95.

Gonçalves, J. F., Resende, M. G., & Mendes, J. J. (2011). A biased random-key genetic algorithm with forward-backward improvement for the resource constrained project scheduling problem. *Journal of Heuristics, 17*(5), 467–486.

Graham, R. L., Lawler, E. L., Lenstra, J. K., & Kan, A. H. G. (1979). Optimization and approximation in deterministic sequencing and scheduling: A survey. *Annals of Discrete Mathematics, 5*, 287–326.

Herroelen, W., De Reyck, B., & Demeulemeester, E. (1998). Resource-constrained project scheduling: A survey of recent developments. *Computers & Operations Research, 25*(4), 279–302.

Hoogeveen, J. A., Lenstra, J. K., & Van de Velde, S. L. (1997). *Sequencing and scheduling: An annotated bibliography*. Eindhoven: Eindhoven University of Technology, Department of Mathematics and Computing Science.

Jackson, J. R. (1955). *Scheduling a production line to minimize maximum tardiness*. Research Report 43, Management Sciences Research Project. Los Angeles: University of California.

Kuo, I. H., Horng, S. J., Kao, T. W., Lin, T. L., Lee, C. L., Terano, T., & Pan, Y. (2009). An efficient flow-shop scheduling algorithm based on a hybrid particle swarm optimization model. *Expert Systems with Applications, 36*(3), 7027–7032.

Lee, K. M., Yamakawa, T., & Lee, K. M. (1998). A genetic algorithm for general machine scheduling problems. In *Knowledge-Based Intelligent Electronic Systems, 1998. Proceedings KES'98. 1998 Second International Conference on* (Vol. 2, pp. 60–66). Piscataway, NJ: IEEE.

Liebchen, C., Schachtebeck, M., Schöbel, A., Stiller, S., & Prigge, A. (2010). Computing delay resistant railway timetables. *Computers and Operations Research, 37*(5), 857–868.

Lin, T. L., Horng, S. J., Kao, T. W., Chen, Y. H., Run, R. S., Chen, R. J., Lai, J. L., & Kuo, I. H. (2010). An efficient job-shop scheduling algorithm based on particle swarm optimization. *Expert Systems with Applications, 37*(3), 2629–2636.

Liu, B., Wang, L., & Jin, Y. H. (2007). An effective PSO-based memetic algorithm for flow shop scheduling. *IEEE Transactions on Systems, Man, and Cybernetics Part B: Cybernetics, 37*(1), 18–27.

Mendes, J. J. D. M., Gonçalves, J. F., & Resende, M. G. (2009). A random key based genetic algorithm for the resource constrained project scheduling problem. *Computers & Operations Research, 36*(1), 92–109.

Nearchou, A. C., & Omirou, S. L. (2006). Differential evolution for sequencing and scheduling optimization. *Journal of Heuristics, 12*(6), 395–411.

Ouelhadj, D., & Petrovic, S. (2009). A survey of dynamic scheduling in manufacturing systems. *Journal of Scheduling, 12*(4), 417–431.

Pezzella, F., Morganti, G., & Ciaschetti, G. (2008). A genetic algorithm for the flexible job-shop scheduling problem. *Computers & Operations Research, 35*(10), 3202–3212.

Prins, C. (2000). Competitive genetic algorithms for the open-shop scheduling problem. *Mathematical Methods of Operations Research, 52*(3), 389–411.

Pruhs, K., Sgall, J., & Torng, E. (2004). Online scheduling. In J. Y.-T. Leung (Ed.), *Handbook of scheduling: Algorithms, models, and performance analysis*. Boca Raton, FL: CRC Press. Chapter 15.

Ribas, I., Leisten, R., & Framiñan, J. M. (2010). Review and classification of hybrid flow shop scheduling problems from a production system and a solutions procedure perspective. *Computers & Operations Research, 37*(8), 1439–1454.

Roshanaei, V., Naderi, B., Jolai, F., & Khalili, M. (2009). A variable neighborhood search for job shop scheduling with set-up times to minimize makespan. *Future Generation Computer Systems, 25*(6), 654–661.

Ruiz, R., & Vázquez-Rodríguez, J. A. (2010). The hybrid flow shop scheduling problem. *European Journal of Operational Research, 205*(1), 1–18.

Taillard, E. (1993). Benchmarks for basic scheduling problems. *European Journal of Operational Research, 64*(2), 278–285.

Williamson, D. P., Hall, L. A., Hoogeveen, J. A., Hurkens, C. A. J., Lenstra, J. K., Sevast'Janov, S. V., & Shmoys, D. B. (1997). Short shop schedules. *Operations Research, 45*(2), 288–294.

Ye, J., & Ma, H. (2015). Multi-objective joint optimization of production scheduling and maintenance planning in the flexible job-shop problem. *Mathematical Problems in Engineering*. doi:10.1155/2015/725460. Accessed March 12, 2016.

Zhang, G., Gao, L., Li, X., & Li, P. (2008). Variable neighborhood genetic algorithm for the flexible job shop scheduling problems. In C. Xiong, Y. Huang, & Y. Xiong (Eds.), *Intelligent Robotics and Applications. First International Conference, ICIRA 2008 Wuhan, China* (pp. 503–512). Berlin/Heidelberg: Springer.

Zhang, G., Gao, L., & Shi, Y. (2010). A genetic algorithm and tabu search for multi objective flexible job shop scheduling problems. In *2010 International Conference on Computing, Control and Industrial Engineering (CCIE)* (Vol. 1, pp. 251–254). Piscataway, NJ: IEEE.

Zhang, C., Li, P., Guan, Z., & Rao, Y. (2007). A tabu search algorithm with a new neighborhood structure for the job shop scheduling problem. *Computers & Operations Research, 34*(11), 3229–3242.

Chapter 6
Location Planning and Network Design

Abstract The planning of a new location belongs to the most complex and difficult planning problems in decision making. Moreover, it belongs to the strategic field of planning which deals with the most important and far-reaching decisions. While the problems considered in the previous chapters are mostly studied on an operative planning level, decisions about single locations or the structure of a supply chain are usually of long-term nature and may be crucial for the future survival and success of a company.

6.1 Location Planning as Multicriteria Decision Problems

When a new facility, e.g. a new warehouse or a new production plant, is to be established, there are usually many (often infinitely many) locations virtually possible. These locations differ in many aspects such as distances to suppliers, customers or other locations, land prices, availability and quality of labor, local regulations and taxes, infrastructure and so on. All these aspects influence the future profitability of the new site and often it is hard to quantify their influence and to estimate the future development of these influencing factors.

Moreover, considering the large number of possible locations it is difficult to obtain all relevant data even for the current situation. In some cases today, geographic information systems (GIS) are applied to substantiate the location decision with the required data (see, e.g., Cheng et al. (2007)). Nevertheless, the provision of sufficiently complete data and the deduction of relevant criteria for a decision remains a major problem in location decisions.

Besides the problem of obtaining the relevant data the multicriteria nature of a decision problem leads to the situation, that there is usually not only one best solution which outperforms other solutions in all criteria. Instead, there is typically a more or less extensive set of solutions denoted as Pareto-optimal or nondominated. For those solutions there does not exist another solution which is better in at least one criterion and not worse in all others. Various methods have been suggested during the last decades which employ additional information from a

T. Hanne, R. Dornberger, *Computational Intelligence in Logistics and Supply Chain Management*, International Series in Operations Research & Management Science 244,
DOI 10.1007/978-3-319-40722-7_6

decision maker to determine a most preferable compromise among those solutions (see, e.g., Hanne 2012). Such preference-based information for solving the decision problem may, for instance, include weights for the criteria, information about desirable outcomes such as reference points, or the assessment of a utility function. In the following we discuss a number of studies on location planning problems which take into account a multicriteria modelling and solution approach.

For instance, Awasthi et al. (2011) consider the following 11 criteria in a model for planning the locations of urban distribution centers: accessibility, security, connectivity to multimodal transport, costs, environmental impact, proximity to customers, proximity to suppliers, resource availability, conformance to sustainable freight regulations, possibility of expansion, and quality of service. In Doerner et al. (2009) only four criteria are considered: the Minisum Facility Location Criterion, the Maximal Covering Location Criterion (MCLC), a risk criterion (the minimization of the tsunami probability), and the overall costs consisting of setup costs of a facility and the additional costs for the required capacity. The minisum objective function measures the sum of distances of clients located within a maximum distance of the assigned facility whereas MCLC counts the number of clients beyond that distance. Both criteria are aggregated to a single one by using a weighted average so that only three criteria remain.

Chen (2001) presents a fuzzy approach applied to an example with five criteria: the investment costs, the expansion possibility, the availability of required material, human resources, and the closeness to the demand market. Another fuzzy multicriteria approach is studied in Chou et al. (2008) who consider 21 criteria for selecting an international tourist hotel location. The criteria relate to aspects concerning the geographical location, traffic conditions, hotel characteristics, and operations management. A fuzzy approach involving traditional multicriteria decision making techniques and a group decision making scenario is analyzed in Chou et al. (2008). In another fuzzy group decision making study Kahraman et al. (2003) consider five evaluation criteria for the selection of facility locations which are proximity to customers, infrastructure, quality of labor, free trade zones, and competitive advantage.

In Harris et al. (2009) three criteria are considered for a facility location problem: minimization of costs, minimization of environmental impact, and minimization of uncovered demand.

A survey on multicriteria location problems is given in Farahani et al. (2010). In total, the study lists 63 publications related to multicriteria location problems but it can be assumed that there are certainly more which are below the detection threshold defined by publications listed on the SCOPUS platform, one of the largest abstract and citation databases for peer-reviewed scientific literature. In the survey the publications are classified into multiattribute problems characterized by a finite, explicitly given set of alternatives and multiobjective problems with an infinite or very large set of alternatives characterized by objective functions and restrictions. The special case of bi-objective problems is considered separately. The found criteria are classified as cost criteria, environmental risks, coverage of the relevant area and equity aspects, service level and effectiveness, profit, and other criteria.

For multiattribute problems, the criteria are often more diverse and may include more versatile aspects related to costs or benefits, available resources, access to public facilities, political issues and regulations, competition aspects, further economic criteria, population aspects, and aspects related to distances.

Despite the obviousness of considering multiple criteria in complex location planning problems, often only a single objective is used in the planning. One reason for that is that usually location problems are already sufficiently complex and hard-to-solve with a single objective. In fact, most of the studied multicriteria variants of location problems are based on the case that there is a rather small finite set of alternatives given whereas situations based either on a possibly infinite set or a very large finite set characterized by various constraints are typical for the single-objective studies.

The other reason is that often the single-objective problems use a complex objective function which aggregates several objectives so that they are not treated explicitly.

In the following sections we will focus on basic variants of location problems which usually assume a single objective function, certainty, and some other often not very realistic assumptions. At the end of the chapter we return to a discussion about using computational intelligence for more complex and realistic variants of location problems.

While most of the considered problems are facility location problems on a single level of a supply chain (e.g. for deciding on a specific number of open distribution centers for delivering customers), others take into account relationships from several levels or are for planning facilities on several levels of a supply chain. In particular, such more complex multi-facility location problems will be treated at the end of the chapter.

For a more thorough treatment of mathematical location theory we refer to Nickel and Puerto (2006). Another excellent overview can be found in Drezner and Hamacher (2001).

6.2 Discrete Location Problems

Mostly, location problems assume that possible locations of facilities are specified in a discrete way, i.e. a finite number of possible locations are considered. Such locations can be assumed as being explicitly given or a graph describing the network structure is specified. In this case it is often assumed that facilities are located at nodes of a graph which leads again to a discrete set under the usual assumptions. Another prerequisite for such problems is information about distances or related costs which can be given explicitly (e.g. in form of a distances matrix) or which can be calculated by a metrics (e.g. based on given coordinates of locations) or using a shortest path approach for a graph-based representation.

6.2.1 The p-Median Problem

6.2.1.1 Problem Statement

In this type of problem p from m possible facility locations are to be selected in order to minimize the total (weighted) distances or transportation costs. All other aspects (e.g. the setup costs for a new facility) are not taken into consideration, i.e. all possible locations are considered as equivalent apart from transportation-related aspects.

The problem can be formulated as follows: Let L be a set of m possible facilities (or location points), C a set n customers (or demand points), and d_{ij}, $i \in \{1,\ldots,n\}$, $j \in \{1,\ldots,m\}$, given distances between facilities and customers. The objective is to determine p locations from L to minimize the sum of these distances (or transportation costs) from the selected locations to the customers:

$$\min \sum_{i=1}^{n} \min_{j \in P} \; d_{ij} \tag{6.1}$$

with $P \subseteq L$ and $|P| = p$.

Using binary decision variables for judging which of the facility locations are used (i.e. belonging to P) and which customers are delivered from which facility we can formulate the following integer optimization problem:

$$\min \sum_{i=1}^{n} \sum_{j=1}^{m} d_{ij} x_{ij} \tag{6.2}$$

such that

$$\sum_{j=1}^{m} x_{ij} = 1, i \in \{1, \ldots, n\} \tag{6.3}$$

$$x_{ij} \leq y_j, i \in \{1, \ldots, n\}, j \in \{1, \ldots, m\} \tag{6.4}$$

$$\sum_{j=1}^{m} y_j = p \tag{6.5}$$

$$x_{ij}, y_j \in \{0, 1\}, i \in \{1, \ldots, n\}, j \in \{1, \ldots, m\} \tag{6.6}$$

$y_j \in \{0, 1\}$ for $j \in \{1, \ldots, m\}$ indicates whether facility j is selected as open ($y_j = 1$) or not ($y_j = 0$, i.e. facility j is closed). $x_{ij} \in \{0, 1\}$ with $i \in \{1, \ldots, n\}$ and $j \in \{1, \ldots, m\}$ indicates whether customer i is served from location j ($x_{ij} = 1$) or not ($x_{ij} = 0$). Constraint (6.3) makes sure that every customer is served by exactly one location,

Eq. (6.4) ensures that only open locations (those belonging to P) may serve a customer. Constraint (6.4) finally guarantees that exactly p locations are open. Note that the objective function could be modified by considering a specific demand a_i of each customer (see, e.g., ReVelle and Swain (1970)) for "weighting" the distances:

$$\min \sum_{i=1}^{n} \sum_{j=1}^{m} a_i d_{ij} x_{ij} \tag{6.7}$$

In Fig. 6.1 an example of a location problem is shown where 15 customers are assigned to 3 facilities selected from a total of 6 possible locations. The p-median problem is proven as to be an NP-hard optimization problem (Kariv and Hakimi 1979). Let us mention that the p-median problem is usually interpreted in the context of facility locations planning, i.e. determining the location of plants, warehouses, or other sites, but it can also be used in other contexts, e.g. for modelling and solving of clustering problems.

Let us note that there are various variants of the p-median problem studied in the literature. For instance, the conditional p-median problem assumes that some of the p facilities to be planned are already determined (Berman and Drezner 2008). Another practically important version of the problem is the capacitated p-median problem which assumes that each facility has a given capacity q_j. In additional constraints it is required that for each facility the assigned demand does not exceed this capacity limit, which can be formulated as in Eq. (6.8).

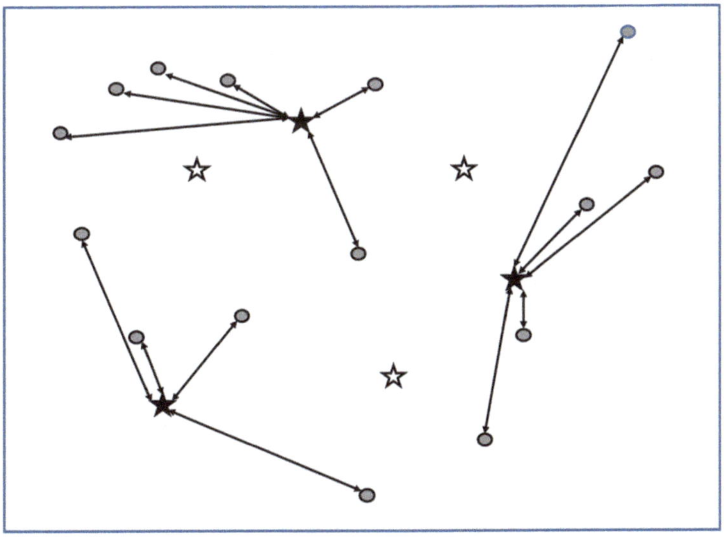

Fig. 6.1 Example of a location problem with 15 customers and 3 from 6 possible facility locations being selected (*open*)

$$\sum_{i=1}^{n} a_i x_{ij} \le q_j y_j, j \in \{1, \ldots, m\} \tag{6.8}$$

Finally there are also stochastic, fuzzy and dynamic versions of the problem investigated in the literature. For instance, a model with fuzzy demand (i.e. fuzzy existence of demand points) is studied in Canós et al. (2001). This allows a better consideration of vague or additional information from reality not present in the standard p-median problem and to calculate solutions which would otherwise not be feasible. This may exhibit a better cost consideration, i.e. obtaining solutions with smaller costs which would not be interpreted as feasible in the standard model.

6.2.1.2 Solution Approaches

Test problems for analyzing and benchmarking algorithms can be found in various libraries. For instance, in Mladenović et al. (2007), six different problem collections are mentioned.

As usual for NP-hard problems we find as possible solution approaches for the p-median problem simple heuristics, metaheuristics, and mathematical approaches. While in the past mathematical approaches dominated for a longer time (with approaches such as Lagrangian relaxation, branch-and-bound, branch-and-cut-and-price, column generation, and dynamic programming), metaheuristic approaches have become increasingly popular during the last 15 years.

However, let us start with considering some simple heuristics. These approaches can be distinguished into construction heuristics and improvement heuristics. Construction heuristics begin, for instance, with a set of just one facility (i.e. a solution of the 1-median problem) and then add step by step additional facilities until p are reached. Each facility is added such that the decrease of costs is maximal. Similarly, it is possible to start with a solution of all m facilities being open and then reducing the number of facilities by one in each iteration such that the cost increase is smallest. This procedure stops when just p facilities are open.

Improvement heuristics or local search approaches can be based on similar considerations for solution modifications. The most popular type of improvement heuristic could be denoted as facility exchange or vertex substitution which considers two, e.g. randomly determined, facilities, one from the open set, the other being closed, and checks whether exchanging them in the considered solution reduces the costs (see, e.g., Chiyoshi and Galvao (2000)).

While the assignment of customers to facilities is rather trivial for the standard p-median problem (since always the closed facility is assigned) it becomes more complex for the capacitated version of the problem. In this case, an exchange heuristic of customers between facilities belongs to the typical improvement steps.

Such local search or improvement heuristics can be used for embedding in other more complex approaches and, for instance, for defining neighborhoods in metaheuristics as used, e.g., in simulated annealing, tabu search, or variable neighborhood search (VNS). For instance, different, increasingly larger neighborhoods in the VNS approach in Hansen and Mladenović (1997) are defined by an increasing number of facilities from the current solution to be exchanged by not yet open facilities (interchange neighborhood structure).

With respect to encoding solutions for a metaheuristics, the following simplification can be used for the standard p-median problem: While there are basically two types of variables in the problem formulation (6.2)–(6.6) it can be sufficient to encode only one during the treatment of solutions in a metaheuristic, that is, the open facilities. The assignment of customers to facilities can be done rather easily by assigning each customer to the nearest (or least costly) open facility (which can be done in O(p)). In particular, this concept reduces the solution space and should increase the performance of the considered solution approach. For instance, in the genetic algorithm used in Alp et al. (2003), solutions are represented by strings of length p containing the indexes of the open facilities. For such kind of solution representations, modified canonical variation operators would work in evolutionary algorithms, e.g. mutations which exchange a facility in the string by a closed facility (not yet included in the string) or 1-point- or multipoint-crossovers. For the crossover operations, it must be taken care that facilities are not represented twice in the resulting string. For instance, in Correa et al. (2004) for two parent solutions these sets of exclusively represented facilities are determined for each parent and then a random number of facilities from these sets are exchanged for the offspring solutions. However, we find more non-standard implementations in the literature. The recombination used in Alp et al. (2003) is based on first building a union of the facilities from two parents and then subsequently reducing their number to p in a greedy way. That means, in each step it is determined which facility closure produces the smallest cost increase. With respect to mutations, this operator was omitted in the study due to insufficient performance which was possibly because the population size was chosen rather high so that an additional diversity generation was not required.

Such a greedy proceeding could be interpreted as a local search embedded in the evolutionary algorithm. A similar idea can be found in Correa et al. (2004) which they call hypermutation: This computationally expensive and rarely applied evolutionary operator checks for a subset of solutions for each open facility to replace it by a best so far closed facility.

With respect to metaheuristics, the following approaches have been studied in the literature on the p-median problem and variants:

- evolutionary algorithms (Alp et al. 2003; Correa et al. 2004)
- path relinking (Díaz and Fernandez 2006)
- scatter search (Díaz and Fernandez 2006)
- simulated annealing (Al-Khedhairi 2008; Chiyoshi and Galvao 2000)
- tabu search (Rolland et al. 1997)
- variable neighborhood search (Hansen and Mladenović 1997)

Also some less well known approaches have been used such as bionomic algorithms which are some kind of memetic algorithms (Maniezzo et al. 1998). We also find a number of other studies on p-median problems combining various heuristic and metaheuristic concepts.

6.2.2 The p-Center Problem

6.2.2.1 Problem Statement

The p-center problem is similar to the p-median problem. Also here a given number of p facilities are to be chosen from a set of m facilities to serve n customers. The objective, however, is not to minimize the sum of distances or related costs but to minimize the maximal distance between any of the customers and the assigned facility.

The p-center problem can be formulated mathematically as follows:

$$\min \max_{i \in C} \min_{j \in P} \; d_{ij} \tag{6.9}$$

with $P \subseteq L$ and $|P| = p$.

As above we assume that L is a set of m possible facilities (or location points) and $P \subseteq L$ is a subset of p facilities selected as open. C is a set n customers (or demand points), and d_{ij}, $i \in \{1, \ldots, n\}$, $j \in \{1, \ldots, m\}$, are given distances between facilities and customers. The objective is to find p locations from L such that the maximum distance from a customer to its closest open facility is minimized. Note that we could multiply d_{ij} by respective weights specific to a customer (e.g. for individual cargo weights or cost factors) without significantly modifying the problem.

When using binary decision variables for judging which of the facility locations are open (i.e. belonging to P) and which customers are delivered from which facility we can formulate the following integer optimization problem:

$$\min z \tag{6.10}$$

such that

$$\sum_{j=1}^{m} x_{ij} = 1, i \in \{1, \ldots, n\} \tag{6.11}$$

$$x_{ij} \leq y_j, i \in \{1, \ldots, n\}, j \in \{1, \ldots, m\} \tag{6.12}$$

$$\sum_{j=1}^{m} y_j = p \tag{6.13}$$

$$z \geq \sum_{j=1}^{m} d_{ij}x_{ij}, i \in \{1, \ldots, n\} \tag{6.14}$$

$$x_{ij}, y_j \in \{0, 1\}, i \in \{1, \ldots, n\}, j \in \{1, \ldots, m\} \tag{6.15}$$

$y_j \in \{0, 1\}$ for $j \in \{1, \ldots, m\}$ indicates whether facility j is open ($y_j = 1$) or closed ($y_j = 0$). $x_{ij} \in \{0, 1\}$ with $i \in \{1, \ldots, n\}$ and $j \in \{1, \ldots, m\}$ indicates whether customer i is served from location j ($x_{ij} = 1$) or not ($x_{ij} = 0$). Constraint (6.11) ensures that every customer is served by one location. Equation (6.12) guarantees that only open locations serve customers. Equation (6.13) makes sure that exactly p locations are open. Equation (6.14) specifies that the auxiliary variable z is larger or equal to distances between customers and their assigned facilities. Together with the objective of minimizing z, this assumes that customers are assigned to their closest open facilities.

Similar to the p-median problem, there exist various variants of the problem such as a capacitated version (with additional capacity constraints for open facilities and their assigned demand), see e.g. Özsoy and Pınar (2006). In Contreras et al. (2012) a variant with a budget constraint is considered. Moreover, the model considers set-up costs of facilities and cost of establishing new links in the graph representation of the assumed network.

Another variant—similar to the conditional p-median problem—is the conditional p-center problem with some of the p locations being pre-specified (Berman and Drezner 2008). In some other problem formulations, the relationships to coverage problems or coverage requirements are studied. For instance, in the problem formulation analyzed in Lim et al. (2004) it is required for locations to cover the demand of at least a pre-specified number of customers. Another variant of the problem assumes that the minimum distance between facilities and customers should be maximized, which is useful for obnoxious facilities (see, e.g., Klein and Kincaid 1994).

6.2.2.2 Solution Approaches

As the p-center problem is similar to the p-median problem also similar approaches can be used for solving it. As the problem is NP-hard as well (Megiddo and Supowit 1984) metaheuristic approaches have gained a noticeable importance during the last few decades.

Generally, we can note that the p-center problem is less often treated by metaheuristics than the p-median problem and mathematical approaches still dominate. One reason may be that the problem in general is less often analyzed than the p-median problem.

Heuristics often make use of some kind of local search and simple improvement operations such as the interchange (or vertex substitution) heuristic. For instance, metaheuristics based on local search such as tabu search or variable neighborhood search can be based on likewise concepts or generalized concepts (Mladenović et al. 2003). Another example is the large-scale local search as discussed in Scaparra et al. (2004). This method is based on a multi-exchange methodology which uses cyclic exchanges of a sequence of customers to capacity-constrained facilities and path exchanges of multiple customers. Evolutionary algorithms or swarm intelligence approaches are applied only rarely to p-center problems and variants. One example is the use of a bee colony optimization approach in Davidović et al. (2011). The bee colony optimization can basically be used for both constructing and improving a solution. The improvement step consists of selecting a randomly specified number of closed locations added to a considered solution with taking into account that the largest distance among all distances between open locations and customers is reduced. After that the same number of open locations is removed from the solution in a greedy way, i.e. choosing solutions which closure increases the objective values least. While the constructive approach works in an unsatisfactory manner, the improvement bee colony optimization performs successfully in comparison to other metaheuristics in terms of solution quality and run time.

Another example of a metaheuristic application is the usage of an adapted harmony search algorithm in Kaveh and Nasr (2011) which performs quite competitively during numerical experiments in comparison to variable neighborhood search, tabu search, and scatter search.

A simulated annealing approach is investigated in Hassin et al. (2003) where it is used together with ordinary local search and with a novel lexicographic local search which considers exchanges of solutions which do not improve the objective values (largest distance) but the number of locations being responsible for that. This lexicographic local search shows improved performance, also when embedded in other heuristic techniques but best results when used with simulated annealing.

6.2.3 The Uncapacitated Facility Location Problem (UFLP)

One obvious drawback of the above formulations of the p-median problem and the p-center problem is that among the many simplifications compared to practical decisions on locations the setup costs of facilities are not taken into consideration. In practice, such setup costs which can easily be around $50 Mio. US$ for the construction of a mid-size warehouse (see, e.g., International, Inc. 2013) are certainly not negligible. Moreover, the specific number of locations, p, is assumed to be given in these problems and not further taken into consideration.

6.2.3.1 Problem Statement

These assumptions are changed with the facility location problems which consider setup costs and do not require a specific number of facilities to be prescribed. Instead, the number of open facilities results from the solution of the optimization problem. In this problem the costs to be minimized consist of a fixed cost component for setting up a facility at a given location and the transportation costs for transporting the goods to the customers.

Let us assume that there are m possible locations for the facilities denoted by the index values $\{1, \ldots, m\}$ and n customers denoted by the index set $\{1, \ldots, n\}$. The transport costs are given by a matrix $C = [c_{ij}]$ with c_{ij} indicating the costs for a transport from location j to customer i. Alternatively, we could also use distances between locations and multiply them by suitable cost factors. The setup costs of a facility are denoted as f_j for $j \in \{1, \ldots, m\}$.

The Uncapacitated Facility Location Problem (UFLP) can then be specified as follows (see, e.g. Ghosh (2003)):

$$\min \sum_{j \in S} f_j + \sum_{i=1}^{n} \min_{j \in S}{}^{j \in S\, j \in S} c_{ij} \tag{6.16}$$

for a set S of facility locations to be determined, $\emptyset \neq S \subseteq \{1, \ldots, m\}$.

An alternative formulation with binary decision variables would be as follows:

$$\min \sum_{j=1}^{m} f_j y_j + \sum_{i=1}^{n} \sum_{j=1}^{m} c_{ij} x_{ij} \tag{6.17}$$

such that

$$\sum_{j=1}^{m} x_{ij} = 1, \quad i \in \{1, \ldots, n\} \tag{6.18}$$

$$x_{ij} \leq y_j, \quad i \in \{1, \ldots, n\},\ j \in \{1, \ldots, m\} \tag{6.19}$$

$$x_{ij} \in \{0, 1\}, y_j \in \{0, 1\}, \quad i \in \{1, \ldots, n\},\ j \in \{1, \ldots, m\} \tag{6.20}$$

The decision variables y_j indicates whether the respective facility is open whereas the x_{ij} variables indicate whether customer i is served by facility j. Equation (6.18) makes sure that every customer is served by exactly one facility, Eq. (6.19) is for guaranteeing that only open facilities serve customers. Equation (6.20) specifies that all decision variables are binary. The objective function (6.17) defines the total costs consisting of setup costs of open facilities plus the transport costs from assigned facilities to the customers.

Note that several variants of the problem are considered in the literature such as problems with additional budget constraints for limiting the set of open facilities (see, e.g. Ghaderi and Jabalameli (2013), Guha and Khuller (1999)), problems with multiple levels (such as stages in a distribution network) (Tcha and Lee 1984), or problems which consider possible outages of facilities (Guha et al. 2003).

6.2.3.2 Solution Approaches

The uncapacitated facility location problem is known to be NP-hard (Cornuéjols et al. 1983) but a number of approximation algorithms with known worst case quality exist. Moreover, the problem can be solved efficiently for some special cases (Jones et al. 1995).

Like the facility location problems considered before, the UFLP can be solved by simple and problem-specific heuristics, by metaheuristics, and by mathematical approaches.

An example of using a simple heuristic is Ghosh (2003) where improvement moves are considered which either add a new facility location which was closed before or which exchange an open and a closed facility. Such local search concepts are embedded in a tabu search and a backtracking approach. In Sun (2006) a tabu search strategy with more simple moves is explored. These moves only consider the adding and dropping of facilities to and from the open set but the approach performs very competitively in a number of numerical experiments. A similar tabu search version with add and drop moves has been studied also in Al-Sultan and Al-Fawzan (1999). Another approach mainly based on local search but introducing randomness by using different starting solutions is studied in Cura (2010). In a number of experiments this approach performed comparable to some other approaches including several tabu search variants and a genetic algorithm despite its simplicity.

Genetic algorithms were applied to the UFLP, for instance in Jaramillo et al. (2002). Solutions can be represented in that approach by having a bitstring only for the facilities for indicating which are open and which are closed. The assignment of customers in not explicitly required as customers can be assigned to the cheapest or closest open facility. Also genetic operators can be applied in a canonical way. However, a specific crossover operator (fitness-based fusion) and mutations based on exchanges of facility locations are suggested for performance reasons. The same encoding for GA solutions is used in Kratica et al. (2001) but different genetic operators are employed. The crossover operator here uses a randomly created mask for the exchange of bits between two parent solutions. The mutation operator uses random bit changes with a frequency (randomly) controlled by a heuristic. Another GA implementation is discussed in Tohyama et al. (2011) which uses the same solution representation but simple one-point crossovers and bit-changing mutations. As an additional mutation operator, a local search is introduced in the approach.

Another metaheuristic approach which can be used for solving the UFLP is particle swarm optimization (PSO) as discussed in Guner and Sevkli (2008). The PSO algorithm which uses normally continuous variables is modified by adding binary variables which specify whether a facility is open or not. These binary variables are determined from usual continuous variables by a suitable mapping (based on truncation and rounding). Besides that a discrete PSO version is analyzed which uses only discrete representations of solutions. This variant also applies variation operators as used in evolutionary algorithms, mutations, one-point and two-point crossovers. Moreover, both PSO versions were studied with embedded local search. While the pure PSO variants were less convincing during a number of numerical tests, the versions with embedded local search performed very successfully also in comparison with a genetic algorithm and evolutionary simulated annealing. In Wang et al. (2008) another PSO algorithm was used in a multi-population variant and utilizing a parallel implementation.

In Marić et al. (2012) three metaheuristics are studied for a special variant of the UFLP, PSO, simulated annealing and a combination of reduced and standard variable neighborhood search denoted as RVNS-VNS. This last approach performs mostly best in a number of computational experiments.

6.2.4 The Capacitated Facility Location Problem (CFLP)

6.2.4.1 Problem Statement

The Capacitated Facility Location Problem (CFLP) basically equals its uncapacitated version (6.17)–(6.20) with an additional constraint concerning the facility capacities which can be formulated as in Eq. (6.8). Note that the problem is sometimes also denoted as Single Source Capacitated Facility Location Problem (SSCFLP) because it assumes that all demand of a customer is satisfied by a single assigned facility. Following this terminology, occasionally authors allow for the CFLP that the x_{ij} variables are not necessarily binary (but nonnegative real values) which allow for solutions with split demand satisfaction among the facilities (multiple sourcing) (see, e.g., Sankaran 2007).

As for the UFLP, the CFLP exists in several variants studied in the literature, e.g., a two-echelon model (with facilities of different levels in the supply chain) (Tragantalerngsak et al. 1997), a multiperiod CFLP and a multicommodity CFLP (see, e.g., Arostegui et al. 2006) as well as problems with several facilities at the same location (Ghiani et al. 2002).

6.2.4.2 Solution Approaches

The NP-hard CFLP has mainly attracted mathematically oriented solution approaches and more or less simple heuristics (see, e.g. Sridharan 1995), but also a number of metaheuristic approaches. For instance, a very large-scale neighborhood search is explored in Ahuja et al. (2004). Neighborhoods for local search are defined here in two ways, first a neighborhood based on exchanges of facilities assigned to customers in a cyclical way, second a neighborhood defined by facility moves, i.e. opening facilities, closing facilities or exchanging closed and open facilities. Compared with a Lagrangian relaxation heuristic, this approach led to comparable results.

In Arostegui et al. (2006) three metaheuristics have been compared for the CFLP (and two problem variants): tabu search, simulated annealing, and a genetic algorithm. While tabu search performed best for the problem under identical run time allowance and simulated annealing being second best, the genetic algorithm could obtain better results for a smaller number of solutions being evaluated (which could indicate a disadvantageous implementation of the GA).

A more complex hybrid metaheuristic is explored in Chen and Ting (2008). This approach uses repeatedly a Lagrangian heuristic and a Multiple Ant Colony System (MACS) in an alternating fashion. The MACS part of the approach uses two types of pheromone information and two solution construction approaches: One type of pheromone is specific to each facility location. This pheromone information is used together with a preferability indicator depending on the ratio of provided capacity and setup costs of a facility for the probabilistic selection of open facilities. The number of facilities to be selected is also determined in a probabilistic way taking into account the total demand in relationship to the average capacity of a facility. The second type of pheromone is specific to customer-facility edges and is used for assigning customers to open facilities. For the probabilistic assignment of facilities the respective pheromone levels are used together with the reciprocal costs of the respective assignment. As the approach does not directly consider feasibility with respect to the capacity constraint, the objective function is modified by a penalty term for punishing capacity violations. Moreover, the approach integrates a local search which is applied only to the best solution of each iteration. This search is based on two heuristics, one of them using reassignments of customers (possibly also to still closed facilities), the other exchanging customers with respect to their assigned facilities.

A tabu search approach for a multi-sourcing version of the CFLP is used in Sun (2012). The tabu search determines a setting of the y_j variables for the facility status (open or closed) by using add and drop operations which maintain the feasibility of solutions. The settings of the x_{ij} variables which are assumed to be real-valued can be determined afterwards by a network algorithm. The tabu search is enhanced by an embedded solution intensification process including path relinking. Another hybrid approach using two different methods for the two subproblems is explored in Harris et al. (2011) for a biobjective problem formulation. The first method is a GA which uses bit-strings for determining the y_j settings while the x_{ij} variable values are obtained by Lagrangian relaxation.

Further metaheuristics applied to the problem include scatter search which is used in Contreras and Díaz (2008) together with GRASP for diversification generation and tabu search for improving considered solutions. A reactive GRASP is used for a stochastic problem variant in Silva and De la Figuera (2007). Another more recent metaheuristic denoted as Kernel Search was applied to the CFLP in Guastaroba and Speranza (2014).

6.3 Continuous Location Problems

In the general case, it appears too restrictive to limit the number of location alternatives to a small, explicitly given set. Theoretically, all geographical locations (e.g. coordinates on a plane or geographic coordinates) could be possible locations, although, of course, most coordinates would result as infeasible when considering various practical aspects. Another possibility to describe possible locations in a more general sense could be to locate them anywhere in graphs which represent, for instance, a road network, i.e. not just at locations from the finite set of nodes of a graph but possibly along edges as well (see, e.g., Sherali and Nordai (1988)).

6.3.1 The Uncapacitated Multi-facility Weber Problem (UMWP)

Weber problems which belong to the most famous problems in location theory are named after the economist Alfred Weber who considered a related location problem for the first time. Occasionally, such problems are also denoted as Fermat-Weber problems, as problems of finding points minimizing the sum of distances from given points were considered before by the mathematician Pierre de Fermat without having a geographic application in mind. Sometimes, the problem is also denoted as (multi-facility) location-allocation problem in the literature.

6.3.1.1 Problem Statement

The basic idea of the location problem is to find one or several points in the plane which minimize the sum of the distance-dependent transportation costs from these points to n given customer destinations.

To be more specific, it is assumed that there is a set of n customers with known locations which are denoted as two-dimensional vectors (coordinates in the plane) $a_i = (a_{i1}, a_{i2})$. Moreover, there is a set of m facilities with locations to be determined and denoted as $x_j = (x_{j1}, x_{j2})$ for $j \in \{1, \ldots, m\}$. Further decision variables describe the assignment of a customer i to a facility j denoted as x_{ij}. The transport costs per unit of distance of customer i are denoted by c_i (which should usually be proportional to the demand quantity) and the distance is calculated by an l_p metrics. Note that the factor c_i is often left away in respective problem formulations so that specific quantities to be shipped are not considered in the objective function. The uncapacitated multi-facility Weber problem (UMWP) can then be formulated as follows:

$$\min \sum_{i=1}^{n} \sum_{j=1}^{m} x_{ij} c_i d(x_j, a_i) \tag{6.21}$$

$$\sum_{j=1}^{m} x_{ij} = 1, i \in \{1, \ldots, n\} \tag{6.22}$$

$$x_{ij} \in \{0, 1\}, \quad a_j \in \mathfrak{R}^2, \quad i \in \{1, \ldots, n\}, j \in \{1, \ldots, m\} \tag{6.23}$$

The objective function (6.21) describes the costs which are calculated by multiplying the distances with a cost parameter c_i assumed to be proportional to the shipped amount. Equation (6.22) makes sure that each customer is assigned to a facility and Eq. (6.23) defines the assignment specific decision variables as binary and the location-specific decision variables as two-dimensional real-valued vectors.

Mostly the distances used in the problem formulation are calculated by using a Euclidean metrics, i.e.

$$d(x_j, a_i) = \sqrt[2]{(\mathrm{x}_{\mathrm{j}1} - \mathrm{a}_{\mathrm{i}1})^2 + (\mathrm{x}_{\mathrm{j}2} - \mathrm{a}_{\mathrm{i}2})^2}, \tag{6.24}$$

but also other metrics are used in the literature (e.g. rectilinear distances).

Often a relaxed version of the problem is considered with $x_{ij} \geq 0$ being real numbers. While the problem above is a mixed integer optimization problem, the relaxed version is a nonlinear optimization problem with linear constraints. The relaxed problem is also denoted as multisource Weber problem while Eqs. (6.21)–(6.23) may also be called single-source Weber problem.

A further simplification of the problem is to assume $m = 1$ which is also denoted as a single facility Weber problem. There are also variants which do not assume a fixed number of facilities, m, but allow this to be optimized as well, e.g. similar to the facility location problem above (see, e.g., Brimberg et al. 2004).

6.3.1.2 Solution Approaches

Let use note that the UMWP is shown to be NP-hard as most of the considered location problems (Megiddo and Supowit 1984). For solving the problem, most methods come either from the mathematical programming approaches or can be described as either simple heuristics or metaheuristics.

Let us explain one of the heuristics in some more detail, because it is used as a background concept in several other approaches: Cooper's (1964) heuristic is based on the observation that the location and the allocation tasks of the uncapacitated Weber problem can be solved separately without difficulty. When the customer allocation is known, the problem consists of solving m single facility Weber problems which can be done efficiently. On the other hand, when locations are defined, the allocation problem can be solved easily by assigning customers to their nearest facility (assuming no capacity or other constraints). The Cooper heuristic which is also denoted as alternate location-allocation method solves these two problems repeatedly until no further improvements occur.

Apart from this approach, there are other heuristics and approaches based on mathematical programming techniques for solving the problem. Also various approaches from the CI field, especially metaheuristics have been suggested for the UMWP: An approach using ideas from tabu search is explored in Gamal and Salhi (2001). Based on a concept similar to Cooper's heuristic they select initial facility locations based on a furthest distance concept which determines new facility locations step by step looking for customer locations as far away as possible from already determined facility locations. For the new locations it is observed that they are not too close to previous customer locations (tabu points). A different tabu search approach is used by Ohlemüller (1997). Tabu search is used here for determining the assignments of customers to facilities (starting with a random solution). Subsequently the location problems of the facilities are solved as single-facility Weber problems and the objective function can be evaluated. After that neighbor solutions defined by single customer re-assignments are explored taking the tabu list into account. For avoiding premature convergence, a diversification procedure is integrated in the method.

The earliest application of evolutionary algorithms to the UMWP seems to be done by Houck et al. (1996). For representing solutions, floating point numbers are used for the real-valued facility coordinates. For restricting their values lower and upper bound can be applied. The EA uses seven different standard variation operators and, moreover, the Cooper heuristic is embedded in the algorithm to improve the solutions and to evaluate the fitness function. Another evolutionary algorithm suggested in Brimberg et al. (2000) uses the same encoding but a different crossover scheme which includes the consideration of a minimal separation distance of offspring solutions to their parents while the mutation operator is dropped in favor of using the Cooper heuristic for solution improvements. A similar approach is applied in Salhi and Gamal (2003) with an additional operator for injecting new random solutions during the run of the algorithm for increasing diversity and avoiding premature convergence. The obtained results are improved in comparison with previous genetic algorithms.

In Brimberg et al. (2006) a variable neighborhood search (VNS) based on a decomposition approach is suggested (Variable Neighborhood Decomposition Search or VNDS) which uses random solutions from the neighborhoods and improves them subsequently by local search. The decomposition is based on selecting a small number k of facilities which are then moved with respect to their coordinates. Such group of k facilities can be built by considering closeness, at random, or by a mixed strategy based on the two others.

Particle swarm optimization is used in Ghaderi et al. (2012) with assuming upper and lower bounds for the facility coordinates which result from the customer coordinates. Feasibility of solutions is achieved by randomly resetting the coordinates in the solution space or by randomly using the coordinates of a customer. The approach is hybridized with several local search heuristics: the Cooper heuristic and two heuristics based on exchanges of facility locations with customer locations.

Hybrid metaheuristics based on two variants of VNS, a genetic algorithm and several local search heuristics are explored in Jabalameli and Ghaderi (2008).

Also neural networks are used occasionally for solving the problem. A modified Kohonen neural network (self-organizing map) is used in Aras et al. (2006) together with a technique denoted as vector quantization for solving an uncapacitated Weber problem. A similar approach can be found in Hsieh and Tien (2004) and Lozano et al. (1998). The Kohonen network is assumed to provide a canonical representation of the problem. It consists of two layers where the input layer corresponds to the two-dimensional demand locations. The output layer consists of m neurons corresponding to the facilities. Each of these nodes is located on a two-dimensional plane with coordinates interpreted as weights. One of the nodes gets activated when input values (i.e. customer coordinates) are closest to the weights. This corresponds to a respective customer-facility assignment. The learning phase for the algorithm tries to reduce the distances between input values and the weights of the respective neuron which corresponds to the goal of locating facilities as close as possible to the served customers.

A comparison of different metaheuristics including tabu search, simulated annealing, variable neighborhood search, an evolutionary algorithm and ant colony optimization led to the following results for uncapacitated Weber problems solved with a generalized cost function (Bischoff and Dächert 2009): While ACO found the best known solutions most often for smaller problem sizes, tabu search and the evolutionary algorithm appeared as best performing for larger size instances. Moreover, the evolutionary algorithm results as least time-consuming for larger problems.

6.3.2 *The Capacitated Multi-facility Weber Problem (CMWP)*

6.3.2.1 Problem Statement

The capacitated multi-facility Weber problem (CMWP) is basically the uncapacitated version with an additional capacity constraint which needs to be matched with specific demand coming from the customers.

For that purpose, the customers are characterized by known demand of d_i for $i \in \{1,..., n\}$ in addition to their location coordinates. The facilities are characterized by given quantities q_j of a considered good for $j \in \{1, \ldots, m\}$. The decision variables are the locations of the facilities denoted as $x_j = (x_{j1}, x_{j2})$ and, in addition, the quantities to be transported from a facility j to a customer i denoted as q_{ij}. The transport costs are given again by c_{ij} per unit of the item and per unit distance where the distance is calculated by a suitable metrics. Note that these parameters are possibly only dependent on the amount of demand of customer i.

Mathematically, the problem can be formulated as follows:

$$\min \sum_{i=1}^{n} \sum_{j=1}^{m} q_{ij} c_{ij} d(x_j, a_i) \tag{6.25}$$

$$\sum_{j=1}^{m} q_{ij} = d_i, i \in \{1, \ldots, n\} \tag{6.26}$$

$$\sum_{i=1}^{n} q_{ij} = q_j, j \in \{1, \ldots, m\} \tag{6.27}$$

$$q_{ij} \geq 0, \quad i \in \{1, \ldots, n\}, j \in \{1, \ldots, m\} \tag{6.28}$$

Equation (6.25) formulates the total cost function to be minimized. Equation (6.26) makes sure that all demand is covered. Equation (6.27) guarantees that all transportations are covered by sufficient amounts available at the source nodes.

Moreover, the model assumes that the given quantities at the locations and the total demand match, i.e.

$$\sum_{i=1}^{n} d_i = \sum_{j=1}^{m} q_j, \tag{6.29}$$

If this is not the case, dummy customers can be introduced in case of excess quantities at the facilities. If the total demand is too high, the problem is infeasible. However, this can easily be found out before starting to solve the problem.

Let us note that the CMWP adds the additional capacity constraint (6.27) to the UMWP. For this type of problem it is in general no longer feasible to satisfy all demand of a node by the closest facilities. Therefore, the problem requires considering explicit amounts q_{ij} of the commodity shipped from facilities to customers instead of assuming unique assignments of customers to facilities. This multi-sourcing assumption implies that Eq. (6.26) is required additionally as well.

Note that there are variants of the problem which require a single-sourcing, i.e. that all demand of a customer is covered by a single facility. Other variants of the CMWP include, for instance, formulations with multiple commodities (Akyüz et al. 2010) or variants where the number of (open) facilities is not specified a priori (Doong et al. 2007). Gong et al. (1995) study a CMWP with obstacles defined as polygons which are forbidden as facility locations and may not be traversed. There are also stochastic and fuzzy problem formulations discussed in the literature. For instance, incertitude with respect to the expected customer demand can be considered in fuzzy models, see, e.g. Zhou and Liu (2007).

6.3.2.2 Solution Approaches

Compared to the UMWP the CMWP is considered less often in the literature and also approaches dealing with CI are used more infrequently. In Manzour-al-Ajdad et al. (2012) simulated annealing is used for assigning customers to facilities. Starting with an initial solution obtained by the Cooper heuristic, simulated annealing is in particular responsible for regaining feasibility. During the run of the simulated annealing process and subsequently various improvement heuristics are used. A similar strategy is used in Öncan (2013) where very large scale neighborhood search is used for the customer assignments while the location determination logic stems from the Cooper heuristic.

A GRASP methodology is used in Luis et al. (2011). The GRASP concept is used for determining the location coordinates of the facilities taking into account only customer coordinates which makes the problem a discrete one. The assignments are then obtained as in Cooper's heuristic. As this heuristic does not consider capacity constraints an additional approach is applied to reassign customers in case of capacity violations. Moreover, a concept of restricted regions is employed to avoid considering previously chosen locations during the search process.

In Gong et al. (1995) a hybrid evolutionary algorithm is used to solve a CMWP variant with polygon-type obstacles. The EA is used for solving the location part of the problem, i.e. for determining the facility locations whereas some other optimization technique is used for assigning customers to locations, i.e. for solving the allocation part. The EA uses representations based on floating point numbers for the facility coordinates. Feasibility of solutions is ensured by repairing a solution: If a facility is located in a forbidden area, the coordinates are moved to the closest vertex of the obstacle polygon. The crossover operator uses intermediate recombinations as known, e.g., from evolution strategies, i.e. offspring locations are defined at random points on the line between the two parent locations. Two types of mutations are considered and selection is based on an elitist concept while also avoiding similar solutions being selected for the next generation.

A similar hybrid approach based on separating the location and the allocation phase is considered in Wen and Iwamura (2008) for a variant of the CMWP which considers the customer demand as fuzzy. For solving a reformulated version of the problem the simplex algorithm is used for solving the allocation problem while a genetic algorithm is used for the location task.

For a model with stochastic demand, a similar hybrid approach is used in Zhou and Liu (2003) which also applies a genetic algorithm for the location part and an embedded simplex algorithm for the assignment part. The stochasticity is considered in the algorithms by repeated simulations.

In Doong et al. (2007) a CMWP without a fixed number of locations and including setup costs for facilities similar to the CFLP is considered. The suggested solution approach is also a hybrid metaheuristic which uses a GA for determining facility locations and a subgradient method for solving a Lagrange relaxation of the allocation problem.

A different approach is considered in Mohammadi et al. (2010) which uses two interleaved genetic algorithms, one for determining the facility coordinates, the other being embedded for solving the allocation problem.

6.4 Location Routing Problems

Generally speaking, the location problems above link the task of strategic location planning to the subsequent operative planning tasks when respective locations are in use by considering transport distances or related costs. However, this link is achieved only on a rather rough level as mostly distances between facilities and customers are assumed to be "given" or are calculated by a standard metrics. Using a metrics like the Euclidean metrics hardly accommodates the situation of using an actual transport system such as a road network. Sometimes other metrics like the rectilinear metrics are assumed to provide a better measure for orthogonal road networks but this situation is frequently not typical in reality. An even harder drawback of the distance concept is that it neglects possibilities of tour building in reality. As discussed in Chap. 3 transportation frequently allows to bundle smaller shipments which are transported along a common optimized route. Figure 6.2 shows an example with 3 from 6 possible locations being selected and related tours for delivering 15 customers. Note that due to the tour optimization different locations appear as optimal in comparison to Fig. 6.1.

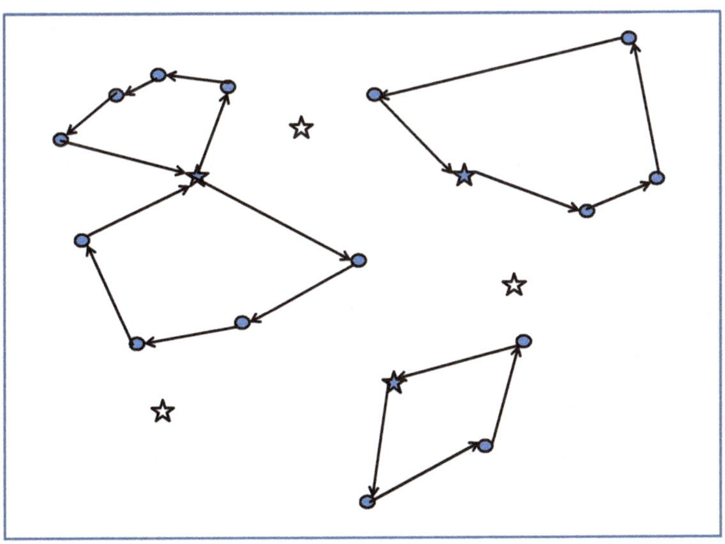

Fig. 6.2 Example of a location routing problem with three from six possible facility locations being selected and respective tours being built

In location routing problems this aspect is taken into consideration by combining location and routing problems. That means that facility locations are to be determined while the considered transportation costs result from solving a respective vehicle routing problem. As discussed in the survey article by Nagy and Salhi (2007) there is a large variety of location routing problems which hardly allows for pointing out a standard type. Loosely speaking, location routing problems can be created from any kind of location problem hybridized with any kind of vehicle routing problem. However, some combinations are more useful from a theoretical or practical point of view.

Most often a capacitated facility location problem is the initial point of formulating a location routing problem, i.e. a scenario with a finite set of possible facility locations with limited capacities for delivering to customers. In the objective, function transports cost and opening costs of facilities are considered. The transport costs result from the solution of a vehicle routing problem assuming specific customer demand to be satisfied. In some variants of the problem the number of facilities may be pre-specified as e.g. for the p-median problem. For a survey on more diverse variants of location routing problem we refer to Drexl and Schneider (2015) where, for instance, dynamic and periodic problems, problems with continuous location decisions, multi-echelon problems, multiobjective problems, arc routing problems, pickup-and-delivery variants, and location routing problems with inventory decisions are considered.

As the basic location problem is mostly a (single-echelon) multi-facility location problem, the underlying vehicle routing problem also has to deal with several depots, i.e. the facilities. The standard location routing problem considered in the survey by Prodhon and Prins (2014) assumes a number of available vehicles (either to be determined or pre-specified) which are homogenous with respect to the capacity, and cause a fixed cost per vehicle used (apart from variable transport costs).

With respect to solution approaches, it appears to be straightforward to solve the two problem parts separately. This would typically assume a hierarchical approach where a solution method is used for the location problem (outer loop) with an embedded second method for solving the vehicle routing problem when routes are to be determined for ascertaining the costs (inner loop). Since already the vehicle routing problem is difficult, it could be assumed that time consumption in such solution approaches may be infeasible. However, if within the outer loop usually only small changes are made, a large part of the vehicle routing problem solution can be maintained. On the other hand, as the planning data for assumed customer demand can be considered uncertain, possibly a fast but less accurate approach for the embedded vehicle routing problem could be sufficient.

In the survey by Nagy and Salhi (2007) metaheuristics where not even discussed as a class of possible solution approaches (although several applications based on variable neighborhood search, tabu search and simulated annealing were mentioned). The situation changed significantly during the following 7 years as the survey by Prodhon and Prins (2014) shows. This survey not only lists more than

20 metaheuristics applications, it also discusses matheuristics which hybridize concepts from mathematical programming with heuristics and metaheuristics.

It is apparent that among the metaheuristic approaches methods based on local search or using neighborhood concepts dominate. This can be explained by the obviousness of using respective approaches which are based on simpler heuristics. Moreover, simplicity and speed of search may be reasons for that. Let us discuss a few examples: For instance, in Derbel et al. (2010) an iterated local search (ILS) is applied to a location routing problem with depot capacities but without vehicle capacities. Starting with an initial solution, the local search only focusses on the vehicle routing part of the problem. Different neighborhood structures are considered, e.g., for intra-route changes or the exchange of customers between different routes. Besides such local search ILS uses a perturbation mechanism which is used in this application for closing and opening facilities with a subsequent reassignment of customers. This ILS concept is integrated in a genetic algorithm in Derbel et al. (2012) where the genetic algorithm uses a solution representation consisting of an assignment part and a route specification part.

In Escobar et al. (2013) we find another example of local search based on a granular tabu strategy. The approach consists of a construction phase and an improvement phase. After constructing an initial solution the facility locations remain fixed so that the improvement phase only considers different local changes inside tours or between tours.

In another study (Jarboui et al. 2013) a location routing problem is solved by a VNS variant with local search (variable neighborhood descent) utilizing five neighborhood structures. The results are compared with those from iterated local search and tabu search where it performs better or equally good as the other two metaheuristics.

In the simulated annealing approach applied in Yu et al. (2010) the method tackles the location and routing problems simultaneously. Solutions are represented as strings which include facilities with their tours and zeros as delimiter symbols. Closed facilities are represented as well with empty tours (i.e. no assigned customers). The simulated annealing approach reverts to local changes such as swap moves, insertion moves, and 2-opt moves. Infeasible solutions are punished within the method by a penalty term.

As discussed in Prodhon and Prins (2014) population-based metaheuristics such as evolutionary algorithms and swarm intelligence approaches are yet of secondary importance for location routing problems, possibly because of less obvious solution representations and adaptations. An interesting exception is the suggestion of a multiple ant colony optimization approach in Ting and Chen (2013). The basic idea of the method is that the location routing problem is divided in three subproblems, the selection of open locations, the assignment of customers to locations, and the vehicle routing problem. The three subproblems are solved in a hierarchical way making use of pheromone exchanges between the different levels. Infeasibility on the customer assignment level is resolved by repair operations based on re-assigning customers or opening additional facilities. Moreover, two approaches for local search are embedded in the solution framework.

6.5 Hub Location Problems

Let us note some aspects of real strategic network design problems which are not considered in the location problems discussed in the previous sections. It could be said that these models only consider one level or stage in a supply chain, e.g. distribution centers which deliver consumers. Real supply chains or supply networks are, however, significantly more complex and include even in simple cases several layers or levels, e.g. suppliers, plants, crossdocking centers, distribution centers, wholesaler, retailers, end customers, some of them repeatedly at different levels of a supply chain.

Moreover, it is interesting to note that even this single decision level is not fully considered. For instance, if the location problems deal with the location of distribution centers which deliver to end customers, it should also be taken into consideration how they are supplied by manufacturing plants. That means, in the objective functions not only the costs for transportation to successive customers should be taken into account but also the transportation costs from upstream facilities such as plants.

Publications which take into account such enriched problem formulations with respect to supply and delivery transportation mostly refer to these problems as hub location problems, e.g. Farahani et al. (2013). Hub locations problems exist in many different variants similar to the variants of the problems considered in the previous sections. For instance, there are different possibilities to specify solutions (discrete or continuous solution domains), different objective functions, different constraints such as capacities, or with respect to the number of hub nodes or whether single or multiple sourcing is assumed.

For solving hub location problems metaheuristics have been applied rather often during recent years. For instance, in the survey study by Farahani et al. (2013), 37 exact approaches based on mathematical programming techniques are listed as being applied to different types of hub location problems. This contrasts with 21 mathematical programming based heuristics, 27 simple or problem-specific heuristics, and 31 approaches based on metaheuristics. The metaheuristic approaches can be subdivided as follows (together with the number of publications):

- tabu search (14 publications)
- genetic algorithms (9 publications)
- simulated annealing (5 publications)
- ant colony optimization (2 publications)
- others (5 publications)

Among the other approaches, all with just one publication, are variable neighborhood search, GRASP, path relinking, and two rather novel metaheuristics denoted as extremal optimization and imperialist competitive algorithm. Note that the numbers of counts are higher than the above number of metaheuristics publications as some studies apply several metaheuristics.

6.6 Multi-Echelon Network Design

Another more general planning and optimization problem would consider the location of facilities at different successive levels in a supply chain, e.g. determining locations of plants and warehouses. Such problems are often considered under the notion of multi-echelon location models (or, as in most cases, two-echelon problems).

In Alumur et al. (2015) a general supply chain model without a specific layer structure is considered: The nodes of the network include arbitrary facilities from a supply chain including customer nodes. Some of these nodes are considered as non-selectable (e.g. customer locations) while others are to be decided with respect to their possible closure. Moreover, a finite number of additional facility locations can be opened. Further decision variables refer to transportation amounts between locations, production amounts and amount of external procurement assuming a multi-product scenario. The relationships between products during the manufacturing are considered by bills of materials. The model assumes limited capacities for production and handling activities with overtime permissibility. The objective function includes set-up and closure costs of facilities, fixed costs during operation, costs for procurement, manufacturing, and transport, overtime costs, and opportunity costs for unsatisfied demand. It is suggested to solve the complex and large-scale model by a general purpose optimization tool.

Another complex model for network design and re-design is considered in Cortinhal et al. (2015). This multi-period model considers four layers with location decisions on two of them (plants and warehouses), limited but expandable capacities, different transportation modes, supplier selection, and outsourcing options. The objective function considers fixed costs and variable costs. Fixed costs are for the opening and closure of a location, for its operation, and for capacity provision. Variable costs include procurement costs for material and optionally finished goods, transport costs and manufacturing costs. The production planning part uses bills of materials to calculate required materials for the production of finished goods.

A recent survey of models for network design and redesign including 18 previous models is given in Cortinhal et al. (2015). Mostly, the solution approaches suggested for complex network design and re-design model are based on mathematical programming approaches. It seems that up to now only in Antunes and Peeters (2001) a CI approach, simulated annealing, is applied to the considered problem. The considered problem can be described as an extension of the capacitated facility location problem which considers two network layers but assumes location decision only on one of them. It can, however, be assumed that respective research in more complex network models including metaheuristics applications will increase in the future.

6.7 Conclusions

As we have seen, there are already a plethora of variants of basic location problems described and analyzed in the literature. Also the usage of CI methods, especially metaheuristics is widespread for these usually NP-hard optimization problems. In general, it is difficult to judge which approach might fit best for which kind of problem. As problem-specific and mostly simple heuristics have played a major rule in the past (and still do) it is not surprising that such concepts are often embedded in more complex metaheuristics for local search. This might also be one of the reasons, why metaheuristics based on local search or using neighborhood concepts (especially tabu search and variable neighborhood search) achieved a leading role among the metaheuristics for solving location problems. Evolutionary algorithms seem to constitute the second largest group of applied CI approaches. A reason for that may be the maturity of these methods but possibly also the flexibility, e.g., to deal with different variable domains (such as binary and real-valued). Swarm intelligence approaches have been used more rarely up-to-now which might have to do with the lower age of many of these approaches. We can therefore assume that future research will reduce this gap. Neural networks only gained some importance for uncapacitated multi-facility Weber problems.

With respect to the considered types of problems we can assert that some of the standard problem types considering facility locations on one level of a supply chain still make up a major part of the research. A majority of the more recent models can be considered as variants of the standard types extending them in one or several aspects. The literature gets sparse when more complex models are considered but it can be assumed that this research direction will expand in the future due to its tremendous practical importance. As mentioned in Drexl and Schneider (2015) there is a "rather general trend in logistics planning towards studying 'richer', more comprehensive and integrated models".

References

Ahuja, R. K., Orlin, J. B., Pallottino, S., Scaparra, M. P., & Scutellà, M. G. (2004). A multi-exchange heuristic for the single-source capacitated facility location problem. *Management Science, 50*(6), 749–760.

Akyüz, M. H., Öncan, T., & Altinel, I. K. (2010). The multi-commodity capacitated multi-facility Weber problem: Heuristics and confidence intervals. *IIE Transactions, 42*(11), 825–841.

Al-Khedhairi, A. (2008). Simulated annealing metaheuristic for solving p-median problem. *International Journal of Contemporary Mathematical Sciences, 3*(28), 1357–1365.

Alp, O., Erkut, E., & Drezner, Z. (2003). An efficient genetic algorithm for the p-median problem. *Annals of Operations Research, 122*(1–4), 21–42.

Al-Sultan, K. S., & Al-Fawzan, M. A. (1999). A tabu search approach to the uncapacitated facility location problem. *Annals of Operations Research, 86*, 91–103.

Alumur, S. A., Kara, B. Y., & Melo, M. T. (2015). Location and logistics. In G. Laporte, S. Nickel, & F. Saldanha da Gama (Eds.), *Location science* (pp. 419–441). Cham: Springer International Publishing.

Antunes, A., & Peeters, D. (2001). On solving complex multi-period location models using simulated annealing. *European Journal of Operational Research, 130*(1), 190–201.

Aras, N., Özkısacık, K. C., & Altınel, İ. K. (2006). Solving the uncapacitated multi-facility Weber problem by vector quantization and self-organizing maps. *Journal of the Operational Research Society, 57*(1), 82–93.

Arostegui, M. A., Kadipasaoglu, S. N., & Khumawala, B. M. (2006). An empirical comparison of tabu search, simulated annealing, and genetic algorithms for facilities location problems. *International Journal of Production Economics, 103*(2), 742–754.

Awasthi, A., Chauhan, S. S., & Goyal, S. K. (2011). A multi-criteria decision making approach for location planning for urban distribution centers under uncertainty. *Mathematical and Computer Modelling, 53*(1), 98–109.

Berman, O., & Drezner, Z. (2008). A new formulation for the conditional p-median and p-center problems. *Operations Research Letters, 36*(4), 481–483.

Bischoff, M., & Dächert, K. (2009). Allocation search methods for a generalized class of location–allocation problems. *European Journal of Operational Research, 192*(3), 793–807.

Brimberg, J., Hansen, P., Mladenovic, N., & Taillard, E. D. (2000). Improvements and comparison of heuristics for solving the uncapacitated multisource Weber problem. *Operations Research, 48*(3), 444–460.

Brimberg, J., Hansen, P., & Mladenović, N. (2006). Decomposition strategies for large-scale continuous location–allocation problems. *IMA Journal of Management Mathematics, 17*(4), 307–316.

Brimberg, J., Mladenovic, N., & Salhi, S. (2004). The multi-source Weber problem with constant opening cost. *Journal of the Operational Research Society, 55*, 640–646.

Canós, M. J., Ivorra, C., & Liern, V. (2001). The fuzzy p-median problem: A global analysis of the solutions. *European Journal of Operational Research, 130*(2), 430–436.

Chen, C. T. (2001). A fuzzy approach to select the location of the distribution center. *Fuzzy Sets and Systems, 118*(1), 65–73.

Chen, C. H., & Ting, C. J. (2008). Combining Lagrangian heuristic and ant colony system to solve the single source capacitated facility location problem. *Transportation Research Part E: Logistics and Transportation Review, 44*(6), 1099–1122.

Cheng, E. W., Li, H., & Yu, L. (2007). A GIS approach to shopping mall location selection. *Building and Environment, 42*(2), 884–892.

Chiyoshi, F., & Galvao, R. D. (2000). A statistical analysis of simulated annealing applied to the p-median problem. *Annals of Operations Research, 96*(1–4), 61–74.

Chou, S. Y., Chang, Y. H., & Shen, C. Y. (2008a). A fuzzy simple additive weighting system under group decision-making for facility location selection with objective/subjective attributes. *European Journal of Operational Research, 189*(1), 132–145.

Chou, T. Y., Hsu, C. L., & Chen, M. C. (2008b). A fuzzy multi-criteria decision model for international tourist hotels location selection. *International Journal of Hospitality Management, 27*(2), 293–301.

Contreras, I. A., & Díaz, J. A. (2008). Scatter search for the single source capacitated facility location problem. *Annals of Operations Research, 157*(1), 73–89.

Contreras, I., Fernández, E., & Reinelt, G. (2012). Minimizing the maximum travel time in a combined model of facility location and network design. *Omega, 40*(6), 847–860.

Cooper, L. (1964). Heuristic methods for location-allocation problems. *SIAM Review, 6*(1), 37–53.

Cornuéjols, G., Nemhauser, G. L., & Wolsey, L. A. (1983). The uncapacitated facility location problem (No. MSRR-493). *Management Sciences Research Report MSRR 493*. Pittsburgh: Carnegie-Mellon University.

Correa, E. S., Steiner, M. T. A., Freitas, A. A., & Carnieri, C. (2004). A genetic algorithm for solving a capacitated p-median problem. *Numerical Algorithms, 35*(2–4), 373–388.

Cortinhal, M. J., Lopes, M. J., & Melo, M. T. (2015). Dynamic design and re-design of multi-echelon, multi-product logistics networks with outsourcing opportunities: A computational study. *Computers & Industrial Engineering, 90*, 118–131.

Cura, T. (2010). A parallel local search approach to solving the uncapacitated warehouse location problem. *Computers & Industrial Engineering, 59*(4), 1000–1009.

Davidović, T., Ramljak, D., Šelmić, M., & Teodorović, D. (2011). Bee colony optimization for the p-center problem. *Computers & Operations Research, 38*(10), 1367–1376.

Derbel, H., Jarboui, B., Hanafi, S., & Chabchoub, H. (2010). An iterated local search for solving a location-routing problem. *Electronic Notes in Discrete Mathematics, 36*, 875–882.

Derbel, H., Jarboui, B., Hanafi, S., & Chabchoub, H. (2012). Genetic algorithm with iterated local search for solving a location-routing problem. *Expert Systems with Applications, 39*(3), 2865–2871.

Díaz, J. A., & Fernandez, E. (2006). Hybrid scatter search and path relinking for the capacitated p-median problem. *European Journal of Operational Research, 169*(2), 570–585.

Doerner, K. F., Gutjahr, W. J., & Nolz, P. C. (2009). Multi-criteria location planning for public facilities in tsunami-prone coastal areas. *OR Spectrum, 31*(3), 651–678.

Doong, S. H., Lai, C. C., & Wu, C. H. (2007). Genetic subgradient method for solving location–allocation problems. *Applied Soft Computing, 7*(1), 373–386.

Drexl, M., & Schneider, M. (2015). A survey of variants and extensions of the location-routing problem. *European Journal of Operational Research, 241*(2), 283–308.

Drezner, Z., & Hamacher, H. W. (Eds.). (2001). *Facility location: Applications and theory*. Berlin/Heidelberg: Springer Science & Business Media.

Escobar, J. W., Linfati, R., & Toth, P. (2013). A two-phase hybrid heuristic algorithm for the capacitated location-routing problem. *Computers & Operations Research, 40*(1), 70–79.

Farahani, R. Z., Hekmatfar, M., Arabani, A. B., & Nikbakhsh, E. (2013). Hub location problems: A review of models, classification, solution techniques, and applications. *Computers & Industrial Engineering, 64*(4), 1096–1109.

Farahani, R. Z., SteadieSeifi, M., & Asgari, N. (2010). Multiple criteria facility location problems: A survey. *Applied Mathematical Modelling, 34*(7), 1689–1709.

Gamal, M. D. H., & Salhi, S. (2001). Constructive heuristics for the uncapacitated continuous location-allocation problem. *Journal of the Operational Research Society, 52*(7), 821–829.

Ghaderi, A., & Jabalameli, M. S. (2013). Modeling the budget-constrained dynamic uncapacitated facility location–network design problem and solving it via two efficient heuristics: A case study of health care. *Mathematical and Computer Modelling, 57*(3), 382–400.

Ghaderi, A., Jabalameli, M. S., Barzinpour, F., & Rahmaniani, R. (2012). An efficient hybrid particle swarm optimization algorithm for solving the uncapacitated continuous location-allocation problem. *Networks and Spatial Economics, 12*(3), 421–439.

Ghiani, G., Guerriero, F., & Musmanno, R. (2002). The capacitated plant location problem with multiple facilities in the same site. *Computers & Operations Research, 29*(13), 1903–1912.

Ghosh, D. (2003). Neighborhood search heuristics for the uncapacitated facility location problem. *European Journal of Operational Research, 150*(1), 150–162.

Gong, D., Gen, M., Xu, W., & Yamazaki, G. (1995). Hybrid evolutionary method for obstacle location-allocation. *Computers & Industrial Engineering, 29*(1), 525–530.

Guastaroba, G., & Speranza, M. G. (2014). A heuristic for BILP problems: The single source capacitated facility location problem. *European Journal of Operational Research, 238*(2), 438–450.

Guha, S., & Khuller, S. (1999). Greedy strikes back: Improved facility location algorithms. *Journal of Algorithms, 31*(1), 228–248.

Guha, S., Meyerson, A., & Munagala, K. (2003). A constant factor approximation algorithm for the fault-tolerant facility location problem. *Journal of Algorithms, 48*(2), 429–440.

Guner, A. R., & Sevkli, M. (2008). A discrete particle swarm optimization algorithm for uncapacitated facility location problem. *Journal of Artificial Evolution and Applications*. doi:10.1155/2008/861512.

Hanne, T. (2012). *Intelligent strategies for meta multiple criteria decision making* (International series in operations research & management science, Vol. 33). New York: Springer Science & Business Media.

Hansen, P., & Mladenović, N. (1997). Variable neighborhood search for the p-median. *Location Science, 5*(4), 207–226.

Harris, I., Mumford, C., & Naim, M. (2009). The multi-objective uncapacitated facility location problem for green logistics. In *IEEE Congress on Evolutionary Computation, 2009. CEC'09* (pp. 2732–2739). Piscataway, NJ: IEEE.

Harris, I., Mumford, C. L., & Naim, M. M. (2011). An evolutionary bi-objective approach to the capacitated facility location problem with cost and CO2 emissions. In N. Krasnogor (Ed.), *Proceedings of the 13th Annual Conference on Genetic and Evolutionary Computation* (pp. 697–704). New York: ACM.

Hassin, R., Levin, A., & Morad, D. (2003). Lexicographic local search and the p-center problem. *European Journal of Operational Research, 151*(2), 265–279.

Houck, C. R., Joines, J. A., & Kay, M. G. (1996). Comparison of genetic algorithms, random restart and two-opt switching for solving large location-allocation problems. *Computers & Operations Research, 23*(6), 587–596.

Hsieh, K. H., & Tien, F. C. (2004). Self-organizing feature maps for solving location–allocation problems with rectilinear distances. *Computers & Operations Research, 31*(7), 1017–1031.

IVI International, Inc. (2013). *IVI construction case study*. 28(10). Accessed March 12, 2016, from http://ivi-intl.com/pdfs/newsletters/2013/IVI_Newsletter_OCT_2013.pdf

Jabalameli, M. S., & Ghaderi, A. (2008). Hybrid algorithms for the uncapacitated continuous location-allocation problem. *The International Journal of Advanced Manufacturing Technology, 37*(1–2), 202–209.

Jaramillo, J. H., Bhadury, J., & Batta, R. (2002). On the use of genetic algorithms to solve location problems. *Computers & Operations Research, 29*(6), 761–779.

Jarboui, B., Derbel, H., Hanafi, S., & Mladenović, N. (2013). Variable neighborhood search for location routing. *Computers & Operations Research, 40*(1), 47–57.

Jones, P. C., Lowe, T. J., Muller, G., Xu, N., Ye, Y., & Zydiak, J. L. (1995). Specially structured uncapacitated facility location problems. *Operations Research, 43*(4), 661–669.

Kahraman, C., Ruan, D., & Doğan, I. (2003). Fuzzy group decision-making for facility location selection. *Information Sciences, 157*, 135–153.

Kariv, O., & Hakimi, S. L. (1979). An algorithmic approach to network location problems. II: The p-medians. *SIAM Journal on Applied Mathematics, 37*(3), 539–560.

Kaveh, A., & Nasr, H. (2011). Solving the conditional and unconditional p-center problem with modified harmony search: A real case study. *Scientia Iranica, 18*(4), 867–877.

Klein, C. M., & Kincaid, R. K. (1994). Technical note—the discrete anti-p-center problem. *Transportation Science, 28*(1), 77–79.

Kratica, J., Tošic, D., Filipović, V., & Ljubić, I. (2001). Solving the simple plant location problem by genetic algorithm. *RAIRO-Operations Research, 35*(1), 127–142.

Lim, A., Rodrigues, B., Wang, F., & Xu, Z. (2004). k-Center problems with minimum coverage. In K. -Y. Chwa & J. I. J. Munro (Eds.), *Computing and Combinatorics, 10th Annual International Conference, COCOON 2004*, Jeju Island, Korea (pp. 349–359). Berlin/Heidelberg: Springer.

Lozano, S., Guerrero, F., Onieva, L., & Larraneta, J. (1998). Kohonen maps for solving a class of location-allocation problems. *European Journal of Operational Research, 108*(1), 106–117.

Luis, M., Salhi, S., & Nagy, G. (2011). A guided reactive GRASP for the capacitated multi-source Weber problem. *Computers & Operations Research, 38*(7), 1014–1024.

Maniezzo, V., Mingozzi, A., & Baldacci, R. (1998). A bionomic approach to the capacitated p-median problem. *Journal of Heuristics, 4*(3), 263–280.

Manzour-al-Ajdad, S. M. H., Torabi, S. A., & Eshghi, K. (2012). Single-source capacitated multi-facility Weber problem—an iterative two phase heuristic algorithm. *Computers & Operations Research, 39*(7), 1465–1476.

Marić, M., Stanimirović, Z., & Milenković, N. (2012). Metaheuristic methods for solving the bilevel uncapacitated facility location problem with clients' preferences. *Electronic Notes in Discrete Mathematics, 39*, 43–50.

Megiddo, N., & Supowit, K. J. (1984). On the complexity of some common geometric location problems. *SIAM Journal on Computing, 13*(1), 182–196.

Mladenović, N., Brimberg, J., Hansen, P., & Moreno-Pérez, J. A. (2007). The p-median problem: A survey of metaheuristic approaches. *European Journal of Operational Research, 179*(3), 927–939.

Mladenović, N., Labbé, M., & Hansen, P. (2003). Solving the p-center problem with tabu search and variable neighborhood search. *Networks, 42*(1), 48–64.

Mohammadi, N., Malek, M. R., & Alesheikh, A. A. (2010). A new GA based solution for capacitated multi source Weber problem. *International Journal of Computational Intelligence Systems, 3*(5), 514–521.

Nagy, G., & Salhi, S. (2007). Location-routing: Issues, models and methods. *European Journal of Operational Research, 177*(2), 649–672.

Nickel, S., & Puerto, J. (2006). *Location theory: A unified approach*. Berlin/Heidelberg: Springer Science & Business Media.

Ohlemüller, M. (1997). Tabu search for large location-allocation problems. *Journal of the Operational Research Society, 48*(7), 745–750.

Öncan, T. (2013). Heuristics for the single source capacitated multi-facility Weber problem. *Computers & Industrial Engineering, 64*(4), 959–971.

Özsoy, F. A., & Pınar, M. Ç. (2006). An exact algorithm for the capacitated vertex p-center problem. *Computers & Operations Research, 33*(5), 1420–1436.

Prodhon, C., & Prins, C. (2014). A survey of recent research on location-routing problems. *European Journal of Operational Research, 238*(1), 1–17.

ReVelle, C. S., & Swain, R. W. (1970). Central facilities location. *Geographical Analysis, 2*(1), 30–42.

Rolland, E., Schilling, D. A., & Current, J. R. (1997). An efficient tabu search procedure for the p-median problem. *European Journal of Operational Research, 96*(2), 329–342.

Salhi, S., & Gamal, M. D. H. (2003). A genetic algorithm based approach for the uncapacitated continuous location–allocation problem. *Annals of Operations Research, 123*(1–4), 203–222.

Sankaran, J. K. (2007). On solving large instances of the capacitated facility location problem. *European Journal of Operational Research, 178*(3), 663–676.

Scaparra, M. P., Pallottino, S., & Scutellà, M. G. (2004). Large-scale local search heuristics for the capacitated vertex p-center problem. *Networks, 43*(4), 241–255.

Sherali, H. D., & Nordai, F. L. (1988). NP-hard, capacitated, balanced p-median problems on a chain graph with a continuum of link demands. *Mathematics of Operations Research, 13*(1), 32–49.

Silva, F. J. F., & De la Figuera, D. S. (2007). A capacitated facility location problem with constrained backlogging probabilities. *International Journal of Production Research, 45*(21), 5117–5134.

Sridharan, R. (1995). The capacitated plant location problem. *European Journal of Operational Research, 87*(2), 203–213.

Sun, M. (2006). Solving the uncapacitated facility location problem using tabu search. *Computers & Operations Research, 33*(9), 2563–2589.

Sun, M. (2012). A tabu search heuristic procedure for the capacitated facility location problem. *Journal of Heuristics, 18*(1), 91–118.

Tcha, D. W., & Lee, B. I. (1984). A branch-and-bound algorithm for the multi-level uncapacitated facility location problem. *European Journal of Operational Research, 18*(1), 35–43.

Ting, C. J., & Chen, C. H. (2013). A multiple ant colony optimization algorithm for the capacitated location routing problem. *International Journal of Production Economics, 141*(1), 34–44.

Tohyama, H., Ida, K., & Matsueda, J. (2011). A genetic algorithm for the uncapacitated facility location problem. *Electronics and Communications in Japan, 94*(5), 47–54.

Tragantalerngsak, S., Holt, J., & Ro, M. (1997). Lagrangian heuristics for the two-echelon, single-source, capacitated facility location problem. *European Journal of Operational Research, 102* (3), 611–625.

Wang, D., Wu, C. H., Ip, A., Wang, D., & Yan, Y. (2008). Parallel multi-population particle swarm optimization algorithm for the uncapacitated facility location problem using openMP. In *IEEE Congress on Evolutionary Computation, 2008. CEC 2008 (IEEE World Congress on Computational Intelligence)* (pp. 1214–1218). Piscataway, NJ: IEEE.

Wen, M., & Iwamura, K. (2008). Facility location–allocation problem in random fuzzy environment: Using (α, β)-cost minimization model under the Hurewicz criterion. *Computers & Mathematics with Applications, 55*(4), 704–713.

Yu, V. F., Lin, S. W., Lee, W., & Ting, C. J. (2010). A simulated annealing heuristic for the capacitated location routing problem. *Computers & Industrial Engineering, 58*(2), 288–299.

Zhou, J., & Liu, B. (2003). New stochastic models for capacitated location-allocation problem. *Computers & Industrial Engineering, 45*(1), 111–125.

Zhou, J., & Liu, B. (2007). Modeling capacitated location–allocation problem with fuzzy demands. *Computers & Industrial Engineering, 53*(3), 454–468.

Chapter 7
Intelligent Software for Logistics

Abstract As we have seen, there are many recent contributions from academics to solving complex logistics problems by advanced methods from computational intelligence and other areas. An obvious question is, of course, how many of them were implemented in practical software solutions and how widespread their use is by companies dealing with these particular problems.

If we consider software tools there are various possibilities to design them. One possibility is to distinguish between software which is mainly focused on optimization algorithms and software which is mainly designed for logistics applications. Optimization software then can be divided into general-purpose optimization software (Sect. 7.1) and tools which focus on specialized methods or particular optimization problems (Sect. 7.2). Logistics software can be divided into general software which usually supports other business processes with optimization features (Sect. 7.3) and software dedicated to logistics in general or specific logistics applications (Sect. 7.4).

7.1 General-Purpose Optimization Software

The usage of mathematical optimization for business problems is normally traced back to the development of the simplex algorithm by George Dantzig in 1947. This method is able to solve linear optimization problems to optimality with a finite number of steps. Although the algorithm may show an exponential complexity, typical optimization problems can be solved with an acceptable number of iterations (and running time). Already in the 1950s some software based on this method has been made available. During the subsequent decades various improvements of this method were suggested and implemented so that today even large-scale problems (with up to several hundred thousand variables and constraints) can be solved in practice. This method (with its improvements) is still competitive and widely used although the so-called interior-point methods have raised much

T. Hanne, R. Dornberger, *Computational Intelligence in Logistics and Supply Chain Management*, International Series in Operations Research & Management Science 244,
DOI 10.1007/978-3-319-40722-7_7

attention since the 1980s because they are able to solve linear optimization problems in polynomial time.

Due to its competitiveness the simplex algorithm (with its improvements) is still the core of optimization software packages today. The supposed market leader in this field is CPLEX Optimization Studio (or just CPLEX) which was developed originally in the 1980s and is provided today by IBM. Other well-known commercial optimization packages are, for instance, XPRESS (Xpress Optimization Suite) provided by FICO, Gurobi (provided by Gurobi Optimization) and MOSEK (provided by MOSEK ApS).

Linear optimization can be used for numerous practice relevant problem types, e.g. in the area of production planning. Referring to logistics problems, however, the variables are often not continuous but integer. Strictly speaking, also in production planning, variables are often integer because products are mostly produced in discrete amounts (i.e. pieces) and not in a quantity expressed by an arbitrary positive real number (as assumed in linear optimization).

Due to the importance of discrete or integer optimization problems relevant algorithms have also been suggested already several decades ago, e.g. cutting planes algorithms since the 1950s, branch & bound since the 1960s, or branch & cut in the 1980s. A more recent development is branch & price which has been used since the 1990s.

Several of these methods have become part of the larger commercial optimization toolboxes. For instance, CPLEX integrated a solver for integer and mixed-integer optimization problems (MIP). The details of that method are not documented but it uses techniques from branching and cutting together with a "bag of tricks" (cf. IBM (2005), Rothberg (n.d.)). These "tricks" include, e.g., strategies for the pre-processing of optimization problems, various ways for using cutting planes, and different heuristics used during the search (e.g. node heuristics) (Lima 2010). Apart from that, CPLEX offers specific support for quadratic optimization problems (continuous and integer/mixed-integer) and a specific solver for constraint programming (CP Optimizer).

Also other packages provide support for integer and mixed-integer problems. XPRESS, for instance, includes a solver called Xpress-MP which is based on branch-and-bound techniques. This solver also applies various types of automatically generated cutting planes and uses heuristics during the search (Laundy et al. 2009). Other types of problems such as quadratic, nonlinear and stochastic programming are also supported by the Xpress software. Moreover, a special solver which is based on constraint programming techniques called Xpress-Kalis is available.

Important as well is the Gurobi software which solves integer or mixed-integer problems also with a branch & bound approach which distinguishes itself by pre-processing techniques, cutting plane generation, the utilization of various heuristics, and the support of parallel processing (Gurobi Optimization 2016).

Apart from these examples there are many more commercial optimization packages available. Wikipedia (2016c), for instance, lists more than forty packages.

In Fourer (2013) even more than fifty packages are listed just for linear programming applications.

In some cases, a respective software package has an even wider focus than just optimization. For instance, general mathematical or statistical software packages often include optimization functionalities although it is not their main focus. Famous examples are Mathematica (www.wolfram.com), MATLAB (www.mathworks.com) or Maple (www.maplesoft.com). MATLAB, for instance, includes optimization approaches in a respective toolbox which provides a considerable number of optimization routines and supports a variety of different problem types.

We can conclude that modern general-purpose optimization toolkits are basically suitable to solve optimization problems as they typically appear in the field of logistics. IBM (n.d.) mentions, for instance, vehicle routing, facility location, scheduling, or network design as possible application areas of the CPLEX mixed-integer optimizer. Nevertheless, three issues should be highlighted when discussing the suitability of this kind of software for intelligent logistics applications: how to set up a suitable model for the optimization software, how to integrate it with logistics applications and how to adapt the method to a problem under consideration.

7.1.1 Setting Up a Suitable Model for the Optimization Software

Setting up a model is a difficult task not only for a beginner in optimization. Since an optimization toolkit is formulated in a programming language, a related problem could theoretically be defined directly in the suitable data structures of the software (e.g. arrays of variables). This would be a demanding task requiring both a good understanding of the programming environment and a good grasp of the mathematical nature of the optimization problem. A more suitable approach would be to formulate problems concisely like in mathematical notations (which then would only require a mathematical understanding). Since mathematical formulations are quite difficult to write with a keyboard (and for some other reasons), a practical approach was found by using a software independent modeling language which allows to formulate a problem similar to a mathematical specification.

An example of such a modeling language is AMPL ("A Mathematical Programming Language", cf. Fourer et al. (2002)) which is based on a mixture of declarative and imperative programming styles. The problem formulation is done mostly through declarative language elements, e.g. sets, decision variables, parameters, objective functions and constraints. Procedural statements can be useful, for instance, for data exchange or the pre- and post-processing of data.

Other common examples of modeling languages are GAMS (General Algebraic Modeling System—www.gams.com), AIMMS (Advanced Interactive Multidimensional Modeling System—business.aimms.com) or OptimJ (Java-based Modeling Language for Optimization—www.ateji.com/optimj).

The use of such a modeling language makes the problem formulation independent from a possible solver. Many software packages, e.g. CPLEX and Xpress, support the problem formulation in such a language. Additionally, some packages come up with their own modeling environment. For instance, CPLEX is equipped with a modeling layer called Concert and a modeling language called OPL (Optimization Programming Language).

7.1.2 Integration of Optimization Software with Logistics Applications

A second problem is how to integrate the optimization software with logistics applications. It is possible to use the optimization software via integrated user interfaces (i.e. for manually entering problem data) but it is obvious that this is not a suitable option for a regular usage of such a tool in a practical setting. On the one hand, this would be much too time-consuming. On the other hand, making a single mistake in the entered data leads to wrong or infeasible solutions or no solution at all due to inconsistencies.

Usually, we can assume that there is already some kind of software for managing logistics activities which should be connected with the optimization software for delivering the data and for redelivering an optimal solution. For instance, a transport service provider operates a transport management system which records transport orders to be carried out,

For the purpose of connection with other applications, commercial toolkits usually provide suitable programming interfaces. For instance, CPLEX provides programming interfaces to C, C++, C#, Java and Python. Moreover, connectors to business applications or mathematical software packages such as Excel or MATLAB are available. Xpress and its modeling layer Mosel provide application programming interfaces to C, C++, Java, .NET and Visual Basic. Gurobi includes interfaces to C, C++, C#, Java, Visual Basic, and Python. In addition, connectors to the mathematical software packages MATLAB and R are available.

So, the basic application programming interfaces are usually available but, of course, if a respective interface in a logistics software does not already exist this implicates substantial additional developing effort for a user. Moreover, when using standard logistics software, also this software must be open for connection with a respective optimization tool.

7.1.3 *Adapting the Method to the Problem Under Consideration*

The problem-specific adaptability is one of the biggest problems for using general-purpose optimization software for advanced logistics planning. Even when it is possible to model a specific logistics problem in the optimization toolkit and to transfer the required data into it, the algorithms can hardly take full advantage of the particularities of the problem.

The usual (mathematical or otherwise) formulations of linear, integer or mixed-integer optimization problems are rather generic. Specific properties of an optimization problem are, therefore, hard to detect and to exploit by a general-purpose optimizer. For some of the standard logistics optimization problems discussed in the previous chapters effort has been spent to model them in a standard way as integer or mixed integer optimization problems to be solvable by standard optimization software. Moreover, in some cases supplementary redundant constraints have been added to increase the efficiency of such software.

A lot of advancements in the optimizers during the last decades was due to efforts in finding and using specific properties of a current problem, e.g. in the pre-processing phase or during the search by utilizing heuristics. Nevertheless, for many types of problems such properties are quite clear when they are analyzed in advance by experts and an adapted algorithm could then deal with them more efficiently.

This is one of the ideas which are usually emphasized when people design metaheuristics. Also metaheuristics can be formulated in a generic way, i.e. not taking particularities of a problem into account, but experience shows that often dramatically better results are obtained when specific data structures (encodings of problems and solutions, modified algorithms, etc.) are used.

Commercial general-purpose solvers are not given this possibility for two reasons: First of all, the source code is usually not accessible and cannot be modified. Secondly, even if a modifiable source code were available, it would be hard to modify it in a suitable way due to the preset basic data structures and the algorithmic concept. In current implementations of general-purpose solvers, a user can mostly only influence an optimization run (and, thus, the performance for the current problem under consideration) by changing some parameters which control the optimization run.

7.2 Software Providing Specific Optimization Algorithms or Supporting Particular Optimization Problems

As we have seen in Sect. 7.1, specific algorithms from computational intelligence do not yet play a significant role in well-known commercial optimization packages. One reason is the historic development of optimization methods and respective software. Commercial packages are still dominated by algorithms having been in use for more than 30 years. Methods of computational intelligence are mostly much newer, at least as far as the awareness of the public is concerned. Another reason is the focus on general-purpose solvers which do not exploit particularities of specific problem types.

Nevertheless, methods from the field of computational intelligence, especially metaheuristics, are widely available at least in academic implementations. When a new method is published, it is usually at least specified by some pseudo code. Moreover, for many methods there is an actual code available which can be freely downloaded from academic websites.

In many cases, the code only provides an implementation of the respective optimization method whereas everything else that might be useful is missing, e.g. codes for data input and output, for pre-processing, visualization, or for the analysis of results. The used data structures and storage formats (if available) are often proprietary and a support for modeling is missing. In particular, the code cannot be directly linked with existing modeling tools and an interfacing with a business application is usually not foreseen.

A better alternative may be software packages which include several of the respective algorithms. They usually provide better support for tasks beyond the pure optimization, e.g. for data input and output, for the visualization of results, or for conducting series of computational experiments. Another advantage is that often several algorithms can be applied to the same type of problem and, thus, one has more security to find a method which is well suited for the problem under consideration.

One example from this class of software packages is ParadisEO (paradiseo. gforge.inria.fr), an object-oriented framework for designing and implementing metaheuristics. The tool comprises several single-solution and population-based metaheuristics, approaches for multiobjective optimization, and it supports parallel and distributed algorithms. It is based on the ANSI-C++ compliant library EO for evolutionary computation and allows for a relatively easy extension by new methods. The package includes tutorials and related documents.

Another example is Hotframe (Heuristic OpTimization FRAMEwork, cf. Fink and Voß (2003)). This software is a C++-based research prototype. It consists of adaptable components including various metaheuristics and descriptions of the collaboration among these components and with applications. Object-oriented classes are used for specifying optimization problems, methods, solutions, neighbors, and solution and move attributes. The implemented methods include, for instance, basic and iterated local search, simulated annealing and similar methods, different variants of tabu search, evolutionary methods, variable depth neighborhood search, candidate list approaches and some hybrid methods.

A third example is the Java-based toolbox OpenOpal (= OPtimization And Learning) (www.openopal.org) which includes especially various metaheuristics (e.g. evolutionary algorithms) and some methods for nonlinear optimization. Moreover, it provides some support of machine learning, data analysis and visualization and is equipped with a graphical user interface for connecting optimization problems and appropriate methods. Let us note that further development of this tool is continued under the names OpenDINO (2016) and OpenCI which is hosted as a Bibucket project.

All these tools are publicly available as free or open-source software. The source code is accessible and can be modified for a better adaptation to a specific problem under consideration. For instance, in OpenOpal and OpenCI, there are generic versions for evolutionary algorithms but also variants adapted to specific problem types.

Apart from these academic implementations, metaheuristics and CI-methods are occasionally also a prominent part of commercial software for optimization. One example is the tool OptQuest (http://www.opttek.com/OptQuest) provided by OptTek Systems, Inc. The approach can be described as a "black box" optimizer which is based on methods such as tabu search, scatter search, integer programming, or neural networks. It provides various interfacing possibilities with other tools, especially simulation software like, for instance, Arena or Taylor ED. Thus, the software can be used for optimizing simulation models which run on another application. Business applications such as Excel can also be integrated with this tool. OptQuest has a callable library written in C, which can be assessed as either a static library or as a dynamic linked library (DLL) or with a COM interface for . NET integration.

Apart from pure optimization software packages, there are non-commercial tools with a wider focus, especially on general mathematical applications—similar to more general commercial software packages as stated above.

One example is Scilab (www.scilab.org), an open source software package for numerical computations which is similar to MATLAB (even regarding the syntax). Its main features and application fields are calculus, linear algebra and analysis, visualization (graphics, animation), time-discrete and continuous simulation and optimization supported by several toolboxes. Scilab includes an interpreter and a high level programming language with interfacing possibilities (e.g. with C, C++, Java, etc.). Similar MATLAB-like but free tools are Octave (http://www.gnu.org/software/octave/) and FreeMat (http://freemat.sourceforge.net/). They also include optimization possibilities but not as extensively as in MATLAB.

Another example is the R software provided by the "R Project for Statistical Computing" (The R Foundation 2016). Although its main focus is on statistical computing and graphics it provides specific packages for optimization infrastructure, continuous optimization and mathematical programming. Also available are interfacing possibilities with other free optimization tools (e.g. Scilab) and commercial optimization software.

Compared with commercial optimization packages several advantages and disadvantages are quite clear: For the free software packages, the source code is usually available and, therefore, a more specific adaptation is possible to particular hard-to-solve optimization problems as they occur in the field of logistics. Moreover, in this group of software, there is a more common usage of methods from computational intelligence and metaheuristics, which have their specific advantages (and some disadvantages, of course as well) compared with the typical algorithms employed in commercial packages. As discussed above, these methods are usually more suitable for problem-specific adaptations than classical optimization algorithms.

An obvious disadvantage of this kind of software is that the software is less mature than in established commercial packages. As the code mostly results from academic projects or voluntary work, these projects are often developed in a less professional way or just simply with too little development effort. Another relating aspect of such projects is that professional support or maintenance is frequently not available. Nevertheless, many open-source development projects have already reached a sufficiently high level of maturity and for some of them commercial support for their utilization in applications may be available.

The last aspect that has to be discussed is the cost issue. Commercial optimization software can be rather expensive. License costs and relating legal issues may be particularly important for a software developing company which wants to integrate an optimization component into their products. Indeed, the cost aspect does not affect open-source software but the terms of the license may impair its usability for specific applications. Frequently, open source licensing schemes such as the GNU General Public License (Wikipedia 2016a) include a "copyleft" condition (Wikipedia 2016b) which permits to use, modify and redistribute the code, but with the limitation that the modified code may not be restricted but has to observe the same licensing conditions as the original software. That may make it difficult (or even impossible) to use a respective code being integrated into some specific application such as a proprietary and commercial logistics software.

7.3 General-Purpose Business Software

Of course, for a company aiming to plan their logistics activities by advanced methods an ideal solution would be to have logistics optimization integrated into the business software which they use on a daily basis. Today, general-purpose business software which supports a large number of processes inside a company is referred to as enterprise resource planning (ERP) software. ERP software is often an integrated suite consisting of several modules for covering different business processes such as procurement, production, sales and order processing.

An ERP software is a transaction-oriented system which manages objects such as cash, materials, and the capacities of resources, and keeps track of the status of these objects.

The planning logic of such software is usually limited. A typical example of planning logic which is available in many standard ERP systems is the Material Requirement Planning (MRP). The purpose of MRP is to determine the materials (parts, items) with respect to quantity and time for covering the demand. If the demand is specified (e.g. by a forecast) the required quantities can be calculated easily.

Given the primary demand for products per period the method uses the bill of materials (BOM) to calculate the needed quantities of pre-products. Other information such as lot sizes for produced quantities or scrap quotas can be taken into account as well. Also time aspects such as production times or delivery times are considered.

One of the main problems of this approach is that the consistency of data is important but often not fully given. For instance, BOMs are incomplete, wrong, or missing. Obviously, this leads to incorrectly calculated quantities with regard to the secondary demand. A main problem of the traditional MRP approach is, however, that the capacities of resources are not taken into account. In particular, for the planned production it is not checked whether the available resource capacities are sufficient. Moreover, alternatives in planning are not considered. For instance, often production can be done by different resources. Also, production, warehousing etc. could possibly take place at different locations. Finally, uncertainties in planning, e.g. with respect to the primary demand, are not included in this planning approach.

Manufacturing Resource Planning (MRP II) is a successor of MRP which considers resource capacities in a simple way. Basically, it works like MRP, but additionally plans are adjusted according to the available capacities of resources. The planning still works successively: i.e. when a first MRP solution leads to results violating the capacities they are subsequently repaired in an arbitrary sequence. Thus, resources are usually not utilized in the best way and optimal solutions cannot be achieved.

The first commercial approaches which tried to extend the planning focus of MRP and MRP II and to use more advanced planning approaches appeared in the 1990s. This software was denoted by the acronym APS which stands for Advanced Planning Systems or Advanced Planning and Scheduling. Another reason for the introduction of APS software was that traditional software from the ERP field was considered to be too slow and not sufficiently user-friendly. A new, more interactive and intelligent planning was considered necessary also taking into account the increased performance of regular computers. Besides that, it should become possible to solve larger planning problems, and planning should not only be done for a single location but also across locations including related logistics activities in an integrated manner. Thus, the planning perspective should be shifted from a site-specific planning towards supply chain management.

Usually, APS software is organized into different modules according to the supply chain planning matrix (see Fig. 7.1). Examples of such modules are network design, demand forecast, distribution & transportation planning, master production planning and production scheduling, or the planning of procurement.

The supply chain planning matrix visualizes different planning fields according to different managerial activities from procurement to distribution, and according to the time horizon from operational towards strategic planning. As logistics comprises cross-sectional activities on different time scales we find related planning tasks at various locations in the supply chain planning matrix. For instance, concrete transportation problems across locations appear during procurement and distribution but possibly also in internal processes if a company consists of several locations. Internal processes include additionally in-house transportations. Warehousing may accompany the procurement (for procured materials), the handling of intermediate products, and the stocking of finished products.

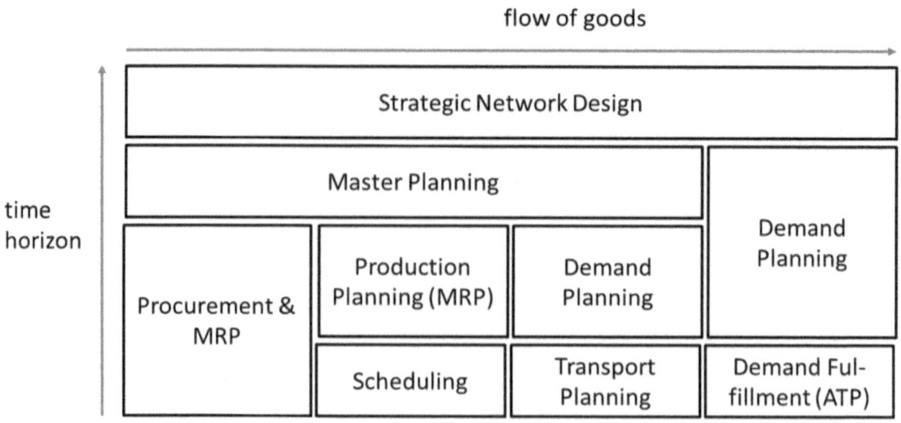

Fig. 7.1 The supply chain planning matrix. SCM related planning activities are characterized by planning horizon and considered processes in the overall flow of goods

Therefore, an APS which supports all related logistics processes could be an ideal solution for a company. Unfortunately, in most cases it is not sufficiently clear whether respective APS software really supports the planning with advanced planning methods. If there is such a support, it usually remains unclear how effective it is.

The market of commercial APS solutions is quite heterogeneous and includes many products provided by smaller companies but also some products provided by software giants like SAP or Oracle. Mostly, the software leaflets, white papers or other information on their company websites describe the features provided by their APS software in a rather enthusiastic way. The underlying techniques, however, usually remain a mystery (cf. Bermudez 1998; Knolmayer 2001). There are two main explanations for this: Either the underlying technologies are kept secret to maintain a competitive advantage, or there are not really many advanced planning methods embedded.

A recent questionnaire-based study (Akabuilo et al. 2011) tried to bring more insight into this issue. It showed that software providers in the APS market mostly include some kind of optimization technique or, at least, some heuristics.

The particular kind of methodology remains unclear in most cases but—if available at all—traditional methods from linear or integer programming seem to dominate. Methods from computational intelligence seem to play a marginal role. For instance, genetic algorithms were used in 11 % of the APS solutions according to software providers who participated in that study. CI methods such as neural networks or fuzzy approaches were not used at all. These basic findings are supported when we look at the support of specific APS modules by advanced planning methods.

Another interesting finding from that study was that often not all of the usually expected modules are included in APS software. In particular, the strongly logistics oriented modules e.g. for distribution & transport planning are only provided by one third of the software companies.

7.4 Logistics Software

The market for logistics software is exceedingly heterogeneous. Usually, there is no single software which supports all logistics activities which might be relevant for a company. Frequently, different logistics activities require different kinds of software. Moreover, there are software solutions which are specific to particular industrial or service sectors. Especially, some types of logistics service providers require their own type of software solutions.

7.4.1 Warehouse Management Systems

Let us start with general software solutions. The heart of in-house logistics is usually called a warehouse management system (WMS) or a warehouse execution system (WES). The foremost purpose of this type of software is the management of inventory and bin locations. Moreover, transfers to stock and releases from stock are to be supported. The respective (in-house) transport processes are to be planned and executed. In many cases, the WMS is connected with subordinate systems for controlling the material flow and related technology (such as Material Flow Control systems or Programmable Logic Controllers).

Other examples of supported activities in in-house logistics are the goods receipt, the order picking, the dispatching of shipments and the stocktaking. Moreover, a WMS supports the protocolling of activities, can generate statistics and reports, and supports communication with other components such as mobile devices (e.g. hand-held scanners), vehicles (e.g. forklift trucks), or with pick-by-light systems.

Typical logistics planning problems such as the production planning, material requirements planning, economic order quantities or external transportation problems are usually not treated by WMS/WES software. It is usually assumed that such planning data are determined otherwise, especially on the level of an ERP system. For a WMS/WES the ERP can be considered as a master system. Operative data such as transportation orders are assumed to come from the ERP level so that the WMS/WES is just refining and executing the respective tasks. Also master data such as article data is assumed to be maintained in the ERP.

Nevertheless, the concrete activities to be executed by the WMS/WES provide fine-level choices and planning potential. For instance, if orders are to be fulfilled, the required material is often available at different locations in the warehouse, e.g. in different aisles. Sometimes, orders can be fulfilled by different suitable resources such as human pickers or automatic storage and retrieval systems (AS/RS). Moreover, there are various possible routes which can be taken during transport (e.g. different conveyor lines).

All these alternatives provide a substantial optimization potential which can be utilized by a WMS/WES or subordinate systems, e.g. material flow control systems. For example, the control system of an AS/RS gets orders that have to be executed (storage and retrieval orders). Their sequence of execution can be varied to save operating time and to increase the performance.

Other decisions are to be made on the higher level of WMS/WES, e.g. the assignment of free bin locations to new bins or the choice of bin locations of required articles. A WMS/WES may also provoke new optional activities such as the relocation of bins. Such relocations may increase the prospective performance of a warehouse if, for instance, items which are assumed to be demanded more frequently in the future are placed in better locations (i.e. locations which cause shorter transportation routes).

Considering specific WMS/WES products we can assert that optimization of warehouse activities is occasionally mentioned but details of used algorithms (or even more detailed descriptions of the optimization problems) are generally missing.

Let us look at a few examples: LMxt is a warehouse management system provided by a-SIS which includes optimization functionalities for several purposes, e.g. the choice of picking locations, pre-cubing (choice of best fitting boxes), or resource allocation (Asis n.d.a). Another product by the same company is called Logys which is said to optimize stocks and automated equipment (Asis n.d.b). Finally, the company offers a tool called LM Transport Order Optimizer for planning and optimizing inter-location transportation by carrier allocation, pre-billing and simulation. Specific algorithms are not mentioned for any of these products.

For another product, the vTradEx eLOG Enterprise Suite WMS (vTradEx n.d.) it is mentioned that "the picking route within the warehouse" is optimized. As a third example, there is the Acteos software. It is said to optimize the resources load, containers, preparation circuits, logistical units for shipping (Acteos n.d.) and to include an inventory optimization strategy and real time optimization methods. However, any kind of details, in particular with respect to underlying algorithms, is missing.

As a final example let us mention the WMS software by Manhattan Associates (n.d.). In this case, it is just remarked that the software "uses advanced algorithms to mathematically organize and optimize warehouse operations".

Compared with APS products (see above), the coverage of advanced optimization techniques by WMS and related software is even more vague. Apart from the usual suspects, business secrets and lack of true optimization techniques, other reasons might be relevant. First of all, the optimization problems might not be as obvious or well-defined as, for instance, in typical transportation problems. Or the optimization benefit may appear to be smaller, since the considered processes are often rather tiny. Finally, WMS products are often customer-specific solutions as in-house logistics systems often lack a sufficient standardization. On the one hand, this might make it more difficult to integrate optimization techniques (which would require a customization as well). On the other hand, due to the partly individual nature of the solutions it might be less likely that details are reported.

7.4.2 *Software for Transportation Planning*

As a next area of logistics software let us reflect on the field of transportation planning. As this is one of the most actively researched fields in the academic literature on logistics optimization we can expect a more mature market for respective commercial software solutions.

A general software which supports the planning, execution, and control of transportation activities is usually called a transportation management system (TMS). Such kind of software may be part of an ERP solution or an APS, or it may work as an add-on to such a system or on a stand-alone basis. Typically, inbound (materials procurement) and outbound (order shipping) activities are supported by such a system but not in-house transportation based on material flow technology (which is usually supported by WMS/WES software).

In particular, a core functionality of a TMS is the determination of routes which are then executed by own resources. Or a suitable service provider is selected. Moreover, there is usually support for the tracing and tracking of the actual transport processes and for subsequent processes such as payment, auditing, and reporting.

The software for transportation planning is regularly surveyed by Hall (2016) and published in OR/MS Today in print and on the journal website. The study is based on questionnaires filled in by respective software vendors (also cf. Partyka and Hall (2000)). The 2016 survey encompasses 25 products offered by 22 companies which responded to the questionnaires. This is a significant increase of the number of products which were 4 years earlier (Hall 2012) just 15 products offered by 12 companies.

From those 25 products, seven use proprietary algorithms which are not specified in more details. One product uses heuristics and constraint programming, seven products use various kinds of metaheuristics and (partly) construction heuristics, seven products use heuristics without a more detailed specification, three products explicitly state to use several heuristics in a combined way and for one product the underlying algorithm was not specified at all. It is interesting to note that 4 years earlier only for three products the usage of (various kinds of) metaheuristics was indicated.

Specific metaheuristics that were mentioned are simulated annealing, genetic algorithms, and local search. In other cases the usage of hybrid or proprietary metaheuristics is specified. Notably, one product from the 2012 survey (IBM ILOG Transportation Analyst) is based on a general optimization product, the IBM ILOG CPLEX Optimization Studio.

All of the products support routing problems for nodes of a network (node routing) while two-thirds also support arc routing, i.e. the visiting of predefined edges of a graph as it is often required for postal delivery or garbage collection. All of the products support a re-routing. All products utilize planning data such as historical travel times or stop times while nine can also utilize real-time traffic information.

Some of the tools provide various interfacing possibilities, e.g. messaging and tracing functions or the integration into order processing systems or some supply chain management software.

In all cases, the driver allocation is supported. In most cases, turn-by-turn navigation, vehicle loading plans and load manifest are supported as well. Pickup and delivery problems are supported by all products. Specific requirements by line transportation service providers like busses are often not supported.

Let us remark that the survey is far from being complete. Other products which are not included in the survey although they include some kind of optimization are, for instance, TransCAD, Transwide, the JDA Transportation Planner or the Voyager Transportation Planning & Management. For instance, the software AccellosOne Optimize (Accellos n.d.) supports transportation planning by load building optimization, scheduling, and routing of vehicles. The website only mentions that the software "utilizes state of the art optimization algorithms to produce loads and routes that balance profitability and customer requirements." Another example is Wide Scope Routing Logistics, a vehicle routing optimization solution provided by the Portuguese company Wide Scope. It supports various vehicle routing problems such as pickup and delivery problems but also includes scheduling and uses various techniques such as mathematical programming, constraint programming and metaheuristics including local search, genetic algorithms, tabu search and ant colony optimization (Wide Scope n.d.).

7.4.3 Packing and Loading Software

Packing and loading problems, e.g. for the loading of vehicles, containers, or pallets, are supported by optimization algorithms in a number of more general logistics software such as commercial transport planning tools (see above). Apart from that there are several specialized tools for solving such problems.

MagicLogic Optimization Inc., for instance, offers a product called "Cube IQ" which addresses in particular the requirements by third-party logistics providers. The used method is described as a BlackBox optimizer based on a "proprietary algorithm developed exclusively by MagicLogic" (Jonker and Smith 2010). Also the consolidation of orders and the loading sequence is taken into account by that tool. The tool can be integrated with WMS and ERP systems.

Another example of loading and packing software products are several tools provided by the Astrokettle group which support cutting stock optimization, rectangle bin packing and scheduling problems (Astrokettle n.d.).

Other software solutions are provided by TOPS software (n.d.). Their tool TOPS pro provides support for optimizing the packaging design, the ship case size or for determining the optimal quantity for an existing ship case. The software includes an interactive carton sizing editor. Another tool called MaxLoad Pro determines how many trucks or containers are needed for a transport and how they should be loaded.

Moreover, this tool helps to determine optimal shipping boxes for an order and to create stable mixed pallets of shipments. The optimization algorithms used in the software are not further specified.

Another tool is the truck, container and pallet loading software by Logen Solutions Corp. (Logen Solutions n.d.). Their tool called CubeMaster supports many different types of loading problems including truck and trailer loading, the loading of sea containers, air containers, pallets, and cartons with various load types (single loads, mixed loads, set, and multi-set loads). Moreover, the software supports various constraints with respect to the loading.

The used optimization algorithm relies on heuristics and metaheuristics based on Tabu Search and a Genetic Algorithm. The approach is said to be published in JORS 2006 but the exact reference was not given and could not be located.

7.5 Conclusions

The market for optimization software appears to be quite mature as many of the general purpose tools have are already been available for several decades. Most of these mature tools rely on established optimization algorithms. These general purpose tools show weaknesses concerning the application to typical logistics problems which essentially result from the computational hardness (NP hardness) of these problems.

Approaches from computational intelligence appear to be suitable approaches to deal with the computational complexity. Although they cannot solve the NP hardness dilemma, they usually lead to good although not optimal solutions for the respective problems.

When new approaches from the area of computational intelligence or metaheuristics are developed, they are usually only available in academic implementations. When they are developed they are sometimes applied to some specific problem in the field of logistics. In other cases, they are demonstrated for some other type of optimization problem and it remains often unclear whether or how they can be used successfully in the area of logistics as well.

Commercial business software mostly provides only limited optimization features (if any). Usually only one (or very few) established optimization methods are implemented. They may work well for the considered types of problems (as they are usually well adapted to the specific application problem) but further improvements resulting from academic research such as new methods are often neglected.

Therefore, there is a huge potential to further improve logistics planning by improved software and it can be assumed that methods from computational intelligence will show a stronger dissemination in logistics software in the future.

References

Accellos. (n.d.). *Transportation optimization*. Accessed March 12, 2016, from http://www.accellos.com/?s=trasnport+optimization

Acteos. (n.d.). *Acteos WMS Warehouse Management System*. Accessed June 6, 2014, from http://www.acteos.com/fr/fr_EN/download/brochure/ACTEOS_WMS_EN.pdf

Akabuilo, E., Dornberger, R., & Hanne, T. (2011). How advanced are advanced planning systems? *Proceedings of the International Symposium on Information Systems and Software Engineering: ISSE 2011*, Orlando, USA, March 27–30, 2011.

Asis. (n.d.a). *LMxt*. Accessed March 12, 2016, from http://www.a-sis.com/en/our-solutions/lmxt-warehouse-and-flow-management-system-large-companies

Asis. (n.d.b). *Logys*. Accessed March 12, 2016, from http://www.a-sis.com/en/our-solutions/logys

Astrokettle. (n.d.). *Astrokettle algorithms*. Accessed March 12, 2016, from http://www.astrokettle.com/index.html

Bermudez, J. (1998). *Advanced planning and scheduling: Is it as goods as it sounds*? (pp. 1–24) The Report on Supply Chain Management (March).

Fink, A., & Voß, S. (2003). HotFrame: A heuristic optimization framework. In S. Voß & D. L. Woodruff (Eds.), *Optimization software class libraries* (pp. 81–154). New York: Springer US.

Fourer, R. (2013). 2013 Linear programming software survey. *OR/MS Today*. Accessed June 6, 2014, from http://www.orms-today.org/surveys/LP/LP-survey.html

Fourer, R., Gay, D. M., & Kernighan, B. W. (2002). *AMPL - A modeling language for mathematical programming* (2nd ed.). AMPL Optimization LLC. Accessed March 12, 2016, from http://ampl.com/resources/the-ampl-book/chapter-downloads/

Gurobi Optimization. (2016). *Mixed integer programming basics*. Accessed March 12, 2016, from http://www.gurobi.com/resources/getting-started/mip-basics

Hall, R. W. (2012). Vehicle routing software survey. *OR/MS Today*. Accessed June 6, 2014, from http://www.orms-today.org/surveys/Vehicle_Routing/vrss.html

Hall, R. W. (2016). Vehicle routing software survey. *OR/MS Today*. Accessed March 12, 2016, from http://www.orms-today.org/surveys/Vehicle_Routing/vrss.html

IBM. (2005). *CPLEX performance tuning for mixed integer programs*. Accessed March 12, 2016, from http://www-01.ibm.com/support/docview.wss?uid=swg21400023

IBM. (n.d.). *Linear programming: An essential optimization technique*. Accessed March 12, 2016, from http://www-01.ibm.com/software/commerce/optimization/linear-programming/

Jonker, R., & Smith, T. (2010). *Load planning software: Cube-IQ BlackBox*. Langley, BC: Magiclogic Optimization Inc. Accessed March 12, 2016, from http://downloads.magiclogic.com/Documents/Cube-IQ_BlackBox_WhitePaper.pdf

Knolmayer, G. (2001). Advanced planning and scheduling systems: Optimierungsmethoden als Entscheidungskriterium für die Beschaffung von Software-Paketen? In U. Wagner (Ed.), *Zum Erkenntnisstand der Betriebswirtschaftslehre am Beginn des 21. Jahrhunderts* (pp. 135–155). Berlin: Duncker & Humblot.

Laundy, R., Perregaard, M., Tavares, G., Tipi, H., & Vazacopoulos, A. (2009). Solving hard mixed-integer programming problems with Xpress-MP: A MIPLIB 2003 case study. *INFORMS Journal on Computing, 21*(2), 304–313.

Lima, R. (2010). IBM ILOG CPLEX What is inside of the box? *EWO Seminar*. Pittsburgh: Carnegie Mellon University.

Logen Solutions. (n.d.). *What's new at Logen Solutions*? Accessed June 6, 2014, from http://www.logensolutions.com/

Manhattan Associates. (n.d.). *Warehouse management: Industry-leading WMS software*. Accessed June 6, 2014, from http://www.manh.com/solutions/distribution-management/warehouse-management

OpenDINO. (2016). *Main page*. Accessed March 12, 2016, from http://www.opendino.org/w/index.php/Main_Page

Partyka, J. G., & Hall, R. W. (2000). On the road to service. *OR/MS Today*.

Rothberg, E. (n.d.). The CPLEX library: Mixed integer programming. *Presentation at 4th Max-Planck Advanced Course on the Foundations of Computer Science,* Saarbrücken. September 8–12, 2003.

The R Foundation. (2016). *The R project for statistical computing*. Accessed March 12, 2016, from http://www.r-project.org/

TOPS Software. (n.d.). *TOPS software solutions for packaging & distribution*. Accessed March 12, 2016, from http://www.topseng.com/

vTradEx. (n.d.). *vTradEx eLOG enterprise suite WMS*. Accessed March 12, 2016, from http://www.vtradex.com/english/solution_4.html

Wide Scope. (n.d.). *Optimization*. Accessed June 6, 2014, from http://routingoptimization.com/optimization.do

Wikipedia. (2016a). *GNU general public license*. Accessed March 12, 2016, from http://en.wikipedia.org/wiki/GNU_General_Public_License

Wikipedia. (2016b). *Copyleft*. Accessed March 12, 2016, from http://en.wikipedia.org/wiki/Copyleft

Wikipedia. (2016c). *List of optimization software*. Accessed March 12, 2016, from http://en.wikipedia.org/wiki/List_of_optimization_software

Authors Brief Biographies

Thomas Hanne received master's degrees in Economics and Computer Science, and a Ph.D. in Economics. From 1999 to 2007 he worked at the Fraunhofer Institute for Industrial Mathematics (ITWM) as senior scientist. Since then he is Professor for Information Systems at the University of Applied Sciences and Arts Northwestern Switzerland and Head of Competence Center Systems Engineering since 2012.

Thomas Hanne is author of about 80 journal and conference articles and editor of several journals and special issues. His current research interests include multicriteria decision analysis, evolutionary algorithms, metaheuristics, optimization, simulation, logistics, and supply chain management.

Rolf Dornberger is the head of the Institute for Information Systems, School of Business, University of Applied Sciences and Arts Northwestern Switzerland FHNW (since 2007) and the head of the competence centers New Trends & Innovation (since 2013) and Technology, Organization & People (since 2014) and was head of the competence center Systems Engineering (2006–2010). In 2002, he was appointed associate professor and, in 2003, full professor for Information Systems at the University of Applied Sciences and Arts Northwestern Switzerland FHNW or rather at its predecessor the University of Applied Sciences Solothurn Switzerland. Additionally, he was a part-time lecturer and visiting professor at the University of Stuttgart and the University of Applied Sciences Zurich. Before returning to academy, he worked in industry in different management positions as a consultant, IT officer and senior researcher in different engineering, technology and IT companies in the field of power generation systems and IT solutions for the airline business. He holds a Ph.D. (1998) and a Diploma degree in Aerospace Engineering (1994). His current research interests include computational intelligence, optimization, innovation and technology management, and new trends and innovations.

T. Hanne, R. Dornberger, *Computational Intelligence in Logistics and Supply Chain Management*, International Series in Operations Research & Management Science 244,
DOI 10.1007/978-3-319-40722-7

Index

T. Hanne, R. Dornberger, *Computational Intelligence in Logistics and Supply Chain Management*, International Series in Operations Research & Management Science 244,
DOI 10.1007/978-3-319-40722-7

FSC
www.fsc.org
MIX
Papier | Fördert
gute Waldnutzung
FSC® C083411

Kolibri 360 GmbH
Ferdinand-Jühlke-Straße 7
99095 Erfurt, Deutschland
produktsicherheit@kolibri360.de